Water

Editor: Danielle Lobban

Volume 463

independence
educational publishers

First published by
Independence Educational Publishers
The Studio, High Green
Great Shelford
Cambridge CB22 5EG
England

ISBN-13: 978 1 86168 923 8

Printed in Great Britain

Printed and bound by the Pureprint Group using it's environmental print technology, a low-carbon, low-waste independently audited process that reduces environmental impact using 100 per cent renewable energy, vegetable-oil-based inks and no industrial alcohol. 98 per cent of all dry waste and 95 per cent of press cleaning solvents are recycled. Pureprint are ISO 14001, EMAS and FSC certified and holds the Queen's Award for Enterprise: Sustainable Development.

FSC www.fsc.org MIX Paper | Supporting responsible forestry FSC® C022913

Acknowledgements

The publisher is grateful for permission to reproduce the material in this book. While every care has been taken to trace and acknowledge copyright, the publisher tenders its apology for any accidental infringement or where copyright has proved untraceable. The publisher would be pleased to come to a suitable arrangement in any such case with the rightful owner.

The material reproduced in **issues** books is provided as an educational resource only. The views, opinions and information contained within reprinted material in **issues** books do not necessarily represent those of Independence Educational Publishers and its employees.

Although every effort has been made to ensure that website addresses are correct at time of going to press, Independence Educational Publishers cannot be held responsible for the content of any website mentioned in this book.

Images

Cover image courtesy of iStock. All other images courtesy of Freepik, Pixabay, Pexels, and Unsplash.

Additional acknowledgements

With thanks to the Independence team: Janey Hills, Klaudia Sommer and Jackie Staines.

Danielle Lobban

Cambridge, May 2025

Contents

Chapter 1: The Power of Water

Chapter 2: Dangers of Water

Chapter 3: A World Without Water

Chapter 4: Being Water Wise

Introduction

Water is volume 463 in the **issues** series. The aim of the series is to offer current, diverse information about important issues in our world, from a UK perspective.

About *Water*

Water – essential, powerful, and at risk. This book explores why water is vital for life, how it can become dangerous, and the growing crisis of scarcity. Discover how to use water responsibly, tackle global challenges, and protect our future.

Our sources

Titles in the **issues** series are designed to function as educational resource books, providing a balanced overview of a specific subject.

The information in our books is comprised of facts, articles and opinions from many different sources, including:

- Newspaper reports and opinion pieces
- Website factsheets
- Magazine and journal articles
- Statistics and surveys
- Government reports
- Literature from special interest groups.

A note on critical evaluation

Because the information reprinted here is from a number of different sources, readers should bear in mind the origin of the text and whether the source is likely to have a particular bias when presenting information (or when conducting their research). It is hoped that, as you read about the many aspects of the issues explored in this book, you will critically evaluate the information presented.

It is important that you decide whether you are being presented with facts or opinions. Does the writer give a biased or unbiased report? If an opinion is being expressed, do you agree with the writer? Is there potential bias to the 'facts' or statistics behind an article?

Activities

Throughout this book, you will find a selection of assignments and activities designed to help you engage with the articles you have been reading and to explore your own opinions. Some tasks will take longer than others and there is a mixture of design, writing and research-based activities that you can complete alone or in a group.

Further research

At the end of each article we have listed its source and a website that you can visit if you would like to conduct your own research. Please remember to critically evaluate any sources that you consult and consider whether the information you are viewing is accurate and unbiased.

Issues Online

The **issues** series of books is complemented by our online resource, Issues Online.

On the Issues Online website you will find a wealth of information, covering over 75 topics, to support the PSHE and RSE curriculum.

Why Issues Online?

Researching a topic? Issues Online is the best place to start for...

Librarians

Issues Online is an essential tool for librarians: feel confident you are signposting safe, reliable, user-friendly online resources to students and teaching staff alike. We provide multi-user concurrent access, so no waiting around for another student to finish with a resource. Issues Online also provides FREE downloadable posters for your shelf/wall/table displays.

Teachers

Issues Online is an ideal resource for lesson planning, inspiring lively debate in class, and setting lessons and homework tasks.

Our accessible, engaging content helps deepen students' knowledge, promotes critical thinking, and develops independent learning skills.

Issues Online saves precious preparation time. We wade through the wealth of material on the internet to filter the best quality, most relevant and up-to-date information you need to start exploring a topic.

Our carefully selected, balanced content presents an overview and insight into each topic from a variety of sources and viewpoints.

Students

Issues Online is designed to support your studies in a broad range of topics, particularly social issues relevant to young people today.

There are thousands of articles, statistics and infographs instantly available to help you with homework, research, and assignments.

With 24/7 access using the powerful Algolia search system, you can find relevant information quickly, easily and safely anytime from your laptop, tablet or smartphone, in class or at home.

Visit issuesonline.co.uk to find out more!

issues online
resources for schools, colleges & libraries

Chapter 1

The Power of Water

What is water?

Water is one of the most essential substances on Earth. It covers about 71% of our planet's surface and exists in oceans, rivers, lakes, and even underground. Despite being so abundant, only a small fraction – around 2.5% – is freshwater, which we rely on for drinking, cooking, farming, and sanitation. Most of this freshwater is frozen in glaciers and polar ice caps, leaving less than 1% accessible for human use.

Water is a unique compound made up of two hydrogen atoms and one oxygen atom (H2O). It exists in three states – solid (ice), liquid (water), and gas (water vapour) – and can easily change between these states through melting, freezing, condensation, and evaporation. This ability to shift forms makes water a key player in the Earth's weather systems and natural cycles.

Why water is essential to life

Without water, life as we know it would not exist. All living organisms – from tiny bacteria to large mammals – need water to survive. Water is a crucial part of the human body, making up about 60% of our mass. It acts as a solvent, transporting nutrients and waste products in and out of cells. It also helps regulate body temperature through sweating and respiration, supports digestion, and cushions joints.

Plants, too, depend on water to grow and thrive. They absorb water from the soil through their roots, which helps transport minerals and nutrients up to the leaves. Through a process called photosynthesis, plants use water and sunlight to create energy, producing oxygen as a by-product. Without water, plants would not survive, and the entire food chain would collapse.

Water is also vital for hygiene and sanitation. Drinking clean water is essential for good health, while washing with water helps prevent the spread of diseases. Access to safe water is a basic human right, but unfortunately, millions of people around the world still lack this access, leading to serious health challenges.

Water's role in the environment

Water is not just important for individuals – it shapes ecosystems and influences the climate. Rivers and lakes provide habitats for fish, birds, and other wildlife, while wetlands act as natural filters, trapping pollutants and improving water quality. The ocean, which contains the majority of the Earth's water, influences weather patterns and supports vast marine biodiversity.

Water also plays a key role in the global climate. It absorbs and distributes heat, helping to regulate temperatures across the planet. The water cycle – involving evaporation, condensation, and precipitation – ensures that water moves from the surface to the atmosphere and back again, replenishing water supplies and maintaining ecosystems.

Why we need to value water

Despite its importance, water is often undervalued and mismanaged. Pollution from industries, agriculture, and households can contaminate water sources, while excessive use can deplete supplies. Climate change is also causing shifts in rainfall patterns, leading to both droughts and flooding, which threaten water availability.

Conserving water is essential to ensure there is enough for everyone, now and in the future. Small actions – like reducing waste, fixing leaks, and protecting natural water sources – can make a big difference. Understanding why water matters helps us appreciate and protect this precious resource.

Water is not just a basic need – it is fundamental to life, the environment, and the survival of all living beings. Let's value it, protect it, and ensure it remains available for generations to come.

The importance of hydration

Good hydration is one of the most important aspects of the diet – drinking enough liquids to keep the fluid levels in the body topped up helps to ensure that all bodily functions are able to take place as normal. Here, Johanna Hignett looks at how much we need to drink, what kind of drinks to choose and the signs of dehydration.

Why do we need fluids?

Water in the body is essential for many important processes to take place. From our blood system carrying essential glucose, oxygen and nutrients to cells, to the kidneys getting rid of waste products we no longer want, fluid in the body is vital to allow these to occur. It also lubricates our joints and eyes, helps our digestive system function and keeps our skin healthy.

We are uniquely designed to regulate our own body temperature, particularly in hot conditions, by losing more water through the skin (perspiration), which in turn acts to cool the body helping to maintain a stable temperature. The more water we lose through sweat, the more we need to replace, so those who live (or holiday) in a hot climate, or lose more fluid than normal doing sporting activities will need to drink more fluids more regularly to replace that lost through sweat.

Water levels in the body change as we age, with newborns having a higher body fluid level compared to adults. Elderly people have lower fluid levels still, but hydration is really important at all ages.

Why do we need to keep hydrated?

Fluid is so important in the body that even when levels drop only slightly, we begin to feel the consequences. Low levels of fluid in the body can cause headaches, feelings of dizziness, lethargy, poor concentration and a dry mouth. Over a longer term, dehydration can cause constipation and can be associated with urinary tract infections and the formation of kidney stones. Regular and adequate intakes of fluid can help to address these.

How much do I need?

Adults need to drink around 1.5–2 litres of fluid a day. A typical mug or glass is about 200 millilitres (ml) so this equates to 8-10 drinks a day. Children need slightly less and should aim for around 6-8 drinks a day, but once they reach teenage years their requirements are similar to adults. Don't forget that fluid needs can vary depending on various factors including level of physical activity and climate, so it is best to remember to drink regularly to keep thirst at bay. A few groups (listed in the box below) need to take particular care to make sure their fluid levels are kept topped up and these include

- Children: Often too busy to recognise the signs of thirst!
- Pregnant and breastfeeding ladies: Fluid is critical for baby and breastmilk
- Older people: Often don't drink enough
- Athletes: Need regular fluid top ups as they lose more through sweat

How do I know I'm drinking enough?

Simply waiting for the sensation of thirst is not a good enough sign of a need to drink – by the time we feel thirsty our body is already dehydrated and potentially suffering some of the effects. Equally, simply drinking to eliminate the feeling of thirst does not fully hydrate the body.

The best indicator of good hydration is urine colour, a pale straw coloured urine being a reliable indicator of good hydration. Darker coloured urine is a sure sign that the body needs more fluid. Getting into the habit of drinking regularly is a great way of keeping hydrated.

What drinks count?

Simply, any drink will help to promote hydration. Water from the tap is a great choice, not least because it's readily available and free. Other drinks count towards our total fluid intake too; milk, fruit juice, tea, coffee and soft drinks are all over 85% water and can be included in total fluid intake.

Some drinks contain other nutrients in addition to fluid such as sugars, fats, vitamins and minerals. Fruit juice is one such drink with vitamins, minerals and sugars too. A glass does contribute to fluid intakes and also provides one of the recommended five portions of fruit and vegetables. Don't forget that the recommended portion is 150ml though, and only one glass of fruit juice counts each day. Regularly choosing drinks with higher levels of added sugars can cause dental problems and is also associated with an increased risk of developing Type 2 diabetes, but there are plenty of sugar free choices available today as alternatives. Finally, if you are choosing sugary drinks try to have them at meal times to limit any problems with dental health.

Drinks with added cream and/or full fat milk provide more calories and higher fat intakes are not generally recommended within healthy eating guidelines, so drinks like creamy hot chocolates are best considered to be treats!

There has been confusion about coffee in particular, with some wrongly thinking it does not hydrate the body as it contains caffeine. Although caffeine is a mild diuretic, the fluid provided by a cup of coffee or tea far outweighs the mild dehydrating effect of caffeine and the fluid levels in the body are improved.

Most of the fluid we consume comes from drinks, but around 20–30% comes from foods. Many fruit and vegetables such as courgettes, cucumber, tomatoes and melon are over 90% water and make a valuable contribution to our overall fluid intake.

Sporty moves

An adequate fluid intake is critical for sports people, principally because their losses of fluid through sweat during activity are increased. It is well known that performance in athletes can drop if fluid levels are not adequate. Athletes often choose isotonic sports drinks to replace both fluid and salts lost through sweat. In fact, regular drinking is as much part of their training as their diet and physical strength. However, most of us who enjoy occasional sport, gym classes, or recreational running don't need the more expensive sports drinks and are best choosing water to keep our levels of hydration up. Remember to take a bottle of water with you when you go for a walk, to the gym, for a run or game of football or rounders and drink a little every now and then!

Don't forget

Good hydration is critical to health and keeps our bodies performing at their best, not only in terms of all the processes that go on but also keeping our levels of concentration up. The best advice is to keep a bottle of water close by to top up fluid levels regularly and keep thirst away.

19 August 2019

Questions

1. Why is water essential for our body's functions?
2. What are some signs that indicate a person might be dehydrated?
3. How can the colour of your urine help you determine your hydration status?
4. Why might children and older adults be more susceptible to dehydration?
5. List three types of drinks that contribute to our daily fluid intake.

Create a hydration diary

Track your daily fluid intake for a week.

Note the types of drinks consumed and the approximate amounts.

Reflect on any patterns or areas for improvement.

Why is water important to the body

By Stephen Makinde

Did you know that your body weight is approximately 60% water?

If you think drinking a few glasses of water is enough, then read below to find out the importance of water in your body.

Life cannot exist without water; it can be a cure for many ailments such as headaches, fatigue, and joint pain, to name some. Water plays a vital role in almost every bodily function, from flushing out waste, regulating body temperature and helping brain function. Water makes up nearly 85% of your brain, about 80% of your blood and about 70% of your lean muscle. We can go for weeks without food, but only three days without water!

Facts that you may not know:

- Men have higher body content that women
- Body water content declines with age to roughly 5% in the elderly
- During a three-hour flight, you can lose up to 1.5 litres of body water!

What happens when you do not drink enough water

We are continually losing water throughout the day through urine, sweat and in our breath. Drinking plenty of water ensures the body is fully hydrated.

Lack of water is one of the triggers for daytime fatigue, even as little as a 2% drop in body water can cause short-term memory, difficulty focusing and studies have found some even struggle doing basic math.

Signs of illnesses, such as fever, vomiting or diarrhoea, cause your body to lose additional fluids. In these cases, you should drink more fluids. Bladder infections and urinary tract stones also require increased water intake.

What water does in the body

Water is essential for digestion, nutrient absorption and chemical reactions. One of the main components of saliva, required for breaking down solid food and keeping your mouth healthy.

The carbohydrates and proteins that our bodies use as food are metabolised and transported by water in the bloodstream. Additionally, it aids in circulation in the body; the levels of oxygen in the bloodstream are at the highest when the body is hydrated. The more oxygen the body has available, the more fat it will burn for energy;without oxygen, the body cannot utilise stored fat for energy. Furthermore, increased levels of oxygen will result in higher levels of energy.

Why hydration is so important

When the body excretes water as sweat during physical activity and in hot environments, the body loses water. Sweat keeps your body cool; however, your body temperature can rise if the water is not replenished.

Water helps remove toxins from the body via sweat, urine and bowel movements.Water is essential for healthy stools and avoiding constipation. Kidneys are important for filtering out waste through urination. Adequate water intake helps your kidneys work efficiently.

By suppressing the appetite, water aids the body to metabolise stored fat. Many of us mistake thirst for hunger; studies have shown that a decrease in water intake will cause fat deposits to increase, while an increase in water intake can actually reduce fat deposits. Another study found that just one glass of water can stop the false sensation of 'hunger' in 100% of the participants.

So how much water should we be drinking?

The general advice on hydration is to drink between six to eight glasses of water a day. Caffeinated beverages may be a concern because caffeine acts as a diuretic and can cause increased urination.Coffee, teas, and many soft drinks contain caffeine. Try to drink caffeinated beverages in moderation and focus on consuming more water.

How to ensure you are drinking enough water

- Keep a bottle of water on your desk at work, or fill a jug of water to drink during the day at home. Seeing it is a reminder and this way, you can monitor how much you have drunk as opposed to keeping a score of how many glasses you have had.
- Eat foods that have higher water content like cucumber, watermelon, lettuce, celery, grapes, oranges, tomatoes.
- If drinking water becomes boring, try some variety like spa water or herbal and fruit teas.
- Either add a squeeze of lemon or lime to your water or add a slice of fruits and herbs to water and leave in a jug or bottle overnight to allow the flavours to infuse into the water.
- If cold drinks are not your thing, then add apple slices and a cinnamon stick in hot water, or even ginger and lemon!

13 January 2020

www.perfectbalanceclinic.com

This is what not drinking enough water will do to your health and wellbeing

Staying hydrated is a simple – but incredibly important – task.

By Camilla Foster

Hydration is a very important, but often overlooked, part of your health.

Water is essential for our bodies to function properly and the impact of not drinking enough water extends beyond mere thirst.

Many of us unknowingly fall short of our daily hydration needs.

What's more, the impact of not drinking enough water affects everything from digestion to skin health, energy levels, and even cognitive function.

Here are five critical ways inadequate hydration can harm your health, highlighting why staying properly hydrated should be a key focus for both your physical health and overall wellbeing.

1. Dry skin

Insufficient hydration can lead to dry skin.

'With a lower water intake, our skin can become dehydrated, making us susceptible to dry, itchy, and cracked skin,' explains Dr Kim Goldin, senior GP and health assessment doctor at Nuffield Health. 'This, in turn, can exacerbate an underlying skin condition such as dermatitis or eczema or make us prone to a skin infection.'

2. Gut issues

Failing to provide your body with enough water can undermine gut health.

'Our gut health can be impacted by low hydration levels as the water content in our intestines decreases,' says Goldin. 'This can result in our bowel movements slowing down and lead to stools becoming hard and dry, potentially causing constipation.'

3. Circulation problems

Dizziness could also be a sign of dehydration.

'Low hydration levels can cause issues with our circulatory system, as there is less blood travelling around the body,' highlights Goldin. 'This can result in symptoms such as low blood pressure, light-headedness, and dizziness.

'In more severe cases of dehydration, our blood can thicken, increasing the risk of blood clots and heart attacks.'

4. Impairs cognitive functioning

'As our brain requires good levels of water to function, our cognitive function can be impacted by low hydration levels,' says Goldin. 'This can lead to poor concentration, mood changes (low mood, irritability), a lowered attention span, and short-term memory issues.'

5. Urinary issues

Dehydration is a common cause of urinary tract infections and kidney stones.

'When fluid intake is low, urine becomes more concentrated, increasing the risk of infection,' explains Goldin. 'Regarding kidney stones, there is a direct link between dehydration and the formation of stones in the kidneys.'

Here are some helpful tips to improve your hydration...

Drink throughout the day

'Even though we are told to drink six to eight glasses of water per day to stay well-hydrated, it is more practical to drink frequently throughout the day,' advises Goldin. 'Setting reminders can help reach this goal.'

Add fruit

'Make drinking water more enjoyable by adding a slice of lemon, cucumber, or even fruits such as berries,' suggests Goldin.

Minimise caffeine and alcohol consumption

'Both caffeine and alcohol act as diuretics and increase water loss from the body – as well as moderating our daily caffeine intake and minimising alcohol consumption, we need to increase our water consumption, e.g., having a glass of water with every caffeinated drink,' recommends Goldin.

Consume foods rich in water content

'Eat foods with high water content, such as watermelon, berries, and cucumbers,' advises Goldin.

Drink extra water when sick

'It is important to increase water intake when feeling unwell, especially if experiencing symptoms such as fever or diarrhoea,' says Goldin.

11 March 2025

Is cold water bad for you? What about drinking from the hose or tap? The facts behind 5 water myths

An article from The Conversation.

By Oliver A.H. Jones, Professor of chemistry, RMIT University

We know the importance of staying hydrated, especially in hot weather. But even for something as simple as a drink of water, conflicting advice and urban myths abound.

Is cold water really bad for your health? What about hot water from the tap? And what is 'raw water'? Let's dive in and find out.

Myth 1: Cold water is bad for you

Some recent TikToks have suggested cold water causes health problems by somehow 'contracting blood vessels' and 'restricting digestion'. There is little evidence for this.

While a 2001 study found 51 out of 669 women tested (7.6%) got a headache after drinking cold water, most of them already suffered from migraines and the work hasn't been repeated since.

Cold drinks were shown to cause discomfort in people with achalasia (a rare swallowing disorder) in 2012 but the study only had 12 participants.

For most people, the temperature you drink your water is down to personal preference and circumstances. Cold water after exercise in summer or hot water to relax in winter won't make any difference to your overall health.

Myth 2: You shouldn't drink hot tap water

This belief has a grain of scientific truth behind it. Hot water is generally a better solvent than cold water, so may dissolve metals and minerals from pipes better. Hot water is also often stored in tanks and may be heated and cooled many times. Bacteria and other disease-causing microorganisms tend to grow better in warm water and can build up over time.

It's better to fill your cup from the cold tap and get hot water for drinks from the kettle.

Myth 3: Bottled water is better

While bottled water might be safer in certain parts of the world due to pollution of source water, there is no real advantage to drinking bottled water in Australia and similar countries.

According to University of Queensland researchers, bottled water is not safer than tap water. It may even be tap water. Most people can't tell the difference either. Bottled water usually costs (substantially) more than turning on the tap and is worse for the environment.

What about lead in tap water? This problem hit the headlines after a public health emergency in Flint, Michigan, in the United States. But Flint used lead pipes with a corrosion inhibitor (in this case orthophosphate) to keep lead from dissolving. Then the city switched water sources to one without a corrosion inhibitor. Lead levels rose and a public emergency was declared.

Fortunately, lead pipes haven't been used in Australia since the 1930s. While lead might be present in some old plumbing products, it is unlikely to cause problems.

Myth 4: Raw water is naturally healthier

Some people bypass bottled and tap water, going straight to the source.

The 'raw water' trend emerged a few years ago, encouraging people to drink from rivers, streams and lakes. There is even a website to help you find a local source.

Supporters say our ancestors drank spring water, so we should, too. However, our ancestors also often died from dysentery and cholera and their life expectancy was low.

While it is true even highly treated drinking water can contain low levels of things like microplastics, unless you live somewhere very remote, the risks of drinking untreated water are far higher as it is more likely to contain pollutants from the surrounding area.

Myth 5: It's OK to drink directly from hoses

Tempting as it may be, it's probably best not to drink from the hose when watering the plants. Water might have sat in there, in the warm sun for weeks or more potentially leading to bacterial buildup.

Similarly, while drinking water fountains are generally perfectly safe to use, they can contain a variety of bacteria. It's useful (though not essential) to run them for a few seconds before you start to drink so as to get fresh water through the system rather than what might have been sat there for a while.

We are fortunate to be able to take safe drinking water for granted. Billions of people around the world are not so lucky.

So whether you like it hot or cold, or somewhere in between, feel free to enjoy a glass of water this summer.

Just don't drink it from the hose.

2 January 2024

Why our oceans are important to human life and our planet

The world's oceans are a precious resource, but they're not everlasting, especially if we continue to mistreat them in the way we do. The ocean's true value to our planet and human life can't and shouldn't be underestimated and that's why it's important we act now to protect and save them from the growing levels of pollution.

What do the oceans do for us?

Understanding just how important the oceans are is a key first step in committing to protecting them and doing all you can to ensure they're kept clean and safe for the benefit of us all. Here are just some of the ways the oceans are integral to the planet and human life.

1. It helps us breathe

Our oceans produce over half of the world's oxygen we breathe and captures an estimate of 40% of all C02 produced, which is four times more than the Amazon rainforest. Phytoplankton that live within the sea are responsible for as much as 50% of this oxygen.

2. It helps regulate the Earth's climate and reduces global warming

Oceans cover 70% of the Earth's surface which absorbs more heat from the sun than any area of land. This allows the sea to swallow and contain heat which may overcome us, if it wasn't absorbed by the water. As the sea water evaporates it becomes denser as the salt quantity increases, this causes it to sink into the depths of the ocean bed. It's like our global climate control system that regulates rain and droughts!

3. It's the largest ecosystem on Earth

The oceans are home to an abundance of life which is valuable to the human food chain. There are only estimates of the number of species that live and thrive in the oceans. Covering 70% of the planet's surface and up to 11,000 metres deep, the number of unknown creatures and plants we've yet to discover could reach into the millions.

Pollution is destroying our oceans

The United Nations Environment Programme estimated that every square mile of oceans contains 46,000 pieces of floating plastic and most plastics and waste in the ocean has come from the land. Land-based sources of pollution include everything from agricultural run-off to pesticides and untreated sewage including plastics.

Floating plastics are amongst the most dangerous and threatening materials due to their ability to break up into microscopic pieces and become ingested by marine life and even find their way back into the food chain. Microplastics are found on almost every beach around the world and plastic debris kills as many as a million seabirds every single year, further highlighting just how much damage is being done to the oceans and the life that relies upon them.

Protecting our oceans

Small changes at home and in the workplace can have a positive impact on ocean pollution and the more changes we make, the larger our impact will be. Consider making these changes or updating your company's policies to reflect your commitment to protecting our oceans.

Say no to single-use plastics

Plastics contain toxins no animal should be eating and with this in mind, cutting back on single-use plastics is a no brainer for the modern business or household. There's no real excuse for single-use plastic bags, coffee cups or water bottles any more. You can buy sustainable alternatives and, in the workplace, you can make it company policy to ban all single-use plastics and provide reusable alternatives to incentivise and get staff on board.

Cut carbon

Our oceans absorb a massive amount of heat from the sun, but it also absorbs almost all the excess heat created when burning fossil fuels. This means waters are warmer than they should be and is damaging to marine life, speeds up the rises of sea levels and can influence weather around the world. Cutting your carbon footprint can help minimise the impact of climate change on the ocean, so consider small changes like energy efficient lightbulbs, timers on electronics so they switch off after a period of not being used, adjusting thermostats down and considering carsharing, cycling or public transport for travel when possible.

There's still time to save our oceans

The International Union for Conservation of Nature has praised the seeming buoyancy and resilience of our seas. Humpback whales' numbers have soared since commercial whaling was banned and the organisation's figures found the proportion of species assessed as threatened with extinction dropped from 18% in 2000 to 11.4% in 2019. Awareness is making a difference but there's still a lot more to be done.

Scientists believe we can reverse the damage done and marine life at risk of being wiped out can recover by 2050, if we provide the right support with less fishing, pollution cuts and habitat restoration. The vital role the oceans play in our world means these actions shouldn't be an afterthought, they should be a priority.

The oceans can exist without us. If we humans become extinct, they won't even notice we're gone. We, however, simply cannot exist without our oceans!

www.wave-utilities.co.uk

What is the water cycle?

The water cycle is a continuous process that moves water around the Earth, helping to keep our planet's ecosystems balanced. It begins when the sun heats up water from oceans, lakes, rivers, and even the soil. This heat causes the water to evaporate, turning it from a liquid into a gas called water vapour. Plants also contribute through a process called transpiration, where water is released from leaves into the air.

Once in the atmosphere, the water vapour cools down as it rises. This cooling causes condensation, where the gas turns back into tiny droplets of liquid. These droplets gather together to form clouds. As more and more water accumulates, the clouds become heavier, and the water eventually falls back to the Earth as precipitation. Precipitation can take many forms, such as rain, snow, sleet, or hail, depending on the temperature.

When precipitation reaches the ground, some of it flows over the surface as runoff, collecting in rivers, lakes, and oceans. Other water seeps into the ground, becoming groundwater, which can feed into rivers or be stored in underground reservoirs. This groundwater eventually makes its way back to the surface or directly into bodies of water, continuing the cycle.

The water cycle is essential for life on Earth, as it provides fresh water for drinking, agriculture, and ecosystems. It also helps regulate the climate by distributing heat through the movement of water. This cycle repeats constantly, keeping our environment in balance.

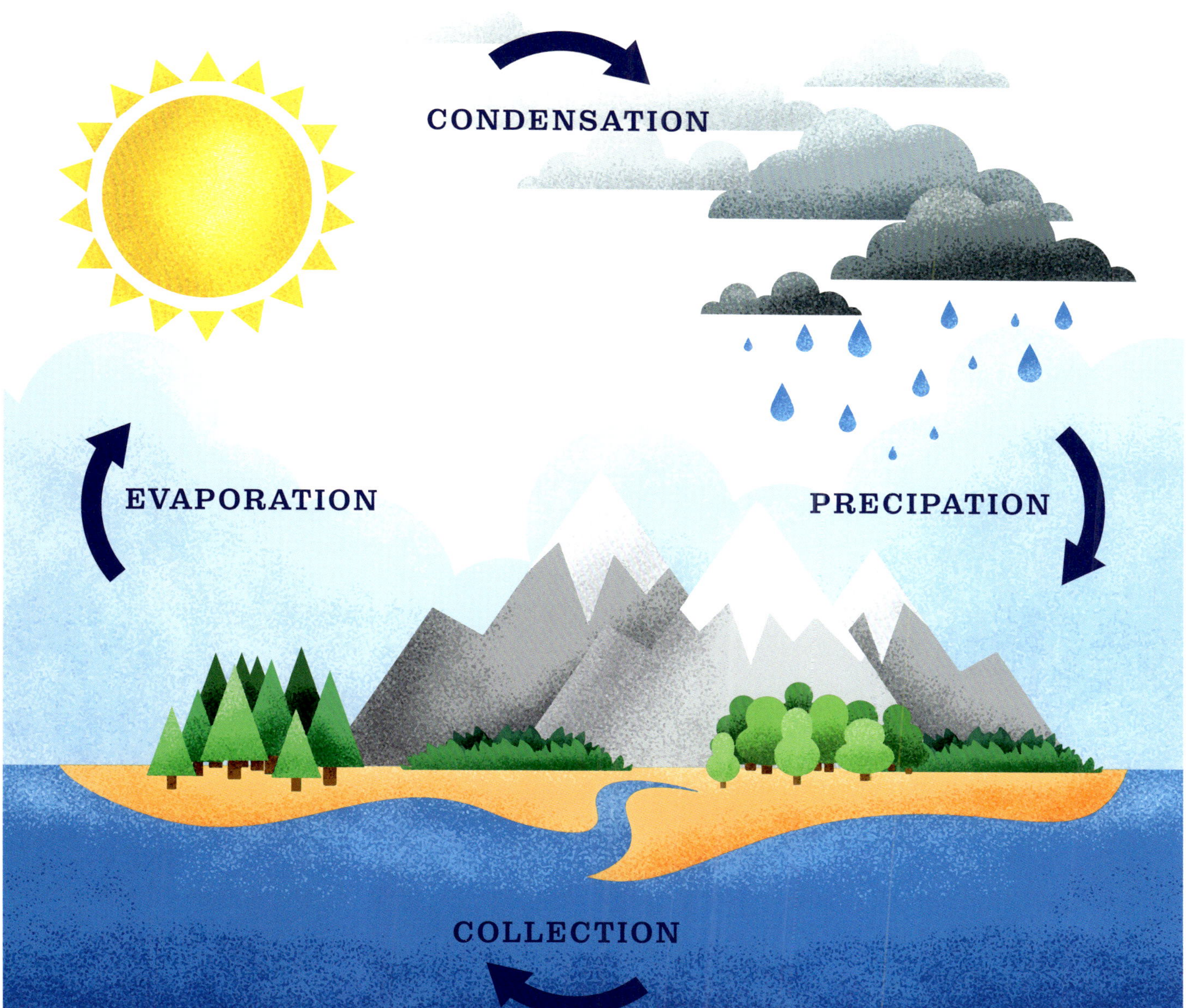

Water: the lifeblood of Earth

Water is often taken for granted, especially in places where it flows easily from taps or falls regularly from the sky. Yet this clear, tasteless liquid is one of the most vital resources on Earth. Without it, life would not exist. From shaping the surface of our planet to keeping our bodies functioning, water plays a central role in supporting all living things. Despite its importance, many people around the world do not have reliable access to safe water. Understanding the role of water – and how we can protect and manage it – is key to safeguarding our future.

Water covers most of the Earth

Approximately 71% of the Earth's surface is covered by water. This includes oceans, rivers, lakes, glaciers, and underground aquifers. The vast majority of this water – around 96.5% – is found in the oceans. These salty waters are not suitable for drinking or for most farming or industrial purposes without expensive treatment.

When people talk about the 'blue planet', they are referring to the fact that water dominates our world. From space, the Earth appears mostly blue due to the huge coverage of its oceans. These bodies of water help regulate the planet's temperature, influence weather patterns, and are home to countless species of marine life. Yet despite the enormous volume of water on Earth, only a tiny portion is available for human use.

A tiny fraction is drinkable

Nearly 97% of the world's water is salty or otherwise undrinkable. That leaves just 3.5% as fresh water. But even then, much of that fresh water is locked away in glaciers, polar ice caps, or deep underground, far from where it can be easily accessed. Only a small portion of fresh water is found in rivers, lakes, and shallow groundwater supplies that can be used for drinking, irrigation, and sanitation.

This makes fresh water a limited and precious resource. It may seem strange to talk about water shortages on a planet that is mostly water, but the reality is that access to clean, fresh water is a major issue in many parts of the world. Pollution, overuse, and climate change all put pressure on the limited supplies we have.

Water is vital to human life

Water is essential to our survival. Up to 60% of the adult human body is made up of water. It is in every cell, tissue, and organ. Water helps regulate body temperature, carries nutrients, removes waste, and cushions joints. It is so important that we cannot survive more than a few days without it.

Humans need between 2.5 and 3.5 litres of water to drink per day, depending on age, activity level, and the climate they live in. This water might come from drinks like tea or juice, or from foods with high water content like fruit and vegetables, but it is vital that we get enough each day.

In addition to drinking, we use water for cooking, washing, flushing toilets, and cleaning. It is also essential in producing the food we eat, generating energy, and manufacturing the products we use. In short, water is woven into every aspect of our daily lives.

Daily water use in the home

In countries like the UK, we use an average of nearly 150 litres of water a day in our homes. That may seem like a lot – and it is. A bath can use around 80 litres, a 10-minute shower about 60 litres, and a single toilet flush up to 13 litres. Even brushing your teeth with the tap running can waste up to 6 litres each time.

Much of this water use goes unnoticed. Yet every drop comes from somewhere and requires energy to treat and transport. Being aware of how we use water – and how we can reduce waste – is an important part of managing this valuable resource.

Access to clean water is not equal

While some people have the luxury of using 150 litres of water a day, others struggle to find even one safe glass. Almost one in ten people on the planet do not have clean water close to home. That means millions of people must walk long distances, often daily, just to collect water for their basic needs. Often, the water they do find is not safe to drink.

Lack of clean water has serious consequences. It affects health, education, and economic development. Diseases like cholera and dysentery spread through dirty water. Children miss school because they are unwell or must help collect water. Adults cannot work or grow food if they are constantly ill.

Improving access to clean water is one of the most effective ways to reduce poverty and improve health around the world. It also helps reduce inequality, especially for women and girls, who are often responsible for collecting water in their communities.

Water and the environment

Water is also essential for the environment. Rivers, wetlands, and lakes provide habitats for wildlife and play a role in filtering water and controlling floods. Healthy ecosystems rely on natural water cycles to stay in balance. When too much water is taken from rivers or when wetlands are drained, the results can be devastating for local plants, animals, and communities.

Climate change is adding more stress to these systems. Rising temperatures and changing rainfall patterns are making droughts more frequent and severe in some areas, while others experience heavier flooding. These extremes make it harder to manage water supplies and protect ecosystems.

Conserving wetlands, protecting rivers, and using water more efficiently are all part of a healthy approach to environmental management. What happens to water in one part of the world can have ripple effects across regions and continents.

Using water responsibly

There are many simple actions individuals can take to use water more responsibly. Turning off the tap while brushing teeth, fixing dripping taps, using water-efficient appliances, and taking shorter showers are just a few ways to reduce household water use. In the garden, collecting rainwater and choosing drought-resistant plants can help save water.

But responsibility also lies with governments, industries, and communities. Investing in water infrastructure, protecting sources of fresh water, and ensuring fair access to safe water are all essential steps. Education plays a key role too – helping people understand why water matters and how they can make a difference.

Conclusion

Water is essential to life on Earth. It covers most of the planet and makes up a major part of our bodies. Yet despite its abundance, only a small fraction is available for us to drink and use. Millions of people live without safe water nearby, while others use it in vast quantities without thinking about where it comes from.

As climate change and population growth increase pressure on water supplies, it is more important than ever to protect this precious resource. That means valuing every drop, using water wisely, and working together to ensure everyone has access to the clean water they need to survive and thrive.

Water connects us all – and safeguarding it is one of the most important challenges we face today. How we respond will shape the future of our planet and the health of generations to come.

When water turns deadly

Water is essential to life. We drink it, bathe in it, and use it in countless ways every day. However, water can also pose serious risks when it becomes contaminated, scarce, or uncontrolled. Understanding these dangers helps us take the necessary precautions to stay safe.

Climate change: a growing threat to water safety

Climate change is having a profound impact on water safety worldwide. As global temperatures rise, weather patterns become more unpredictable, leading to more frequent and severe weather events. This has significant implications for both water availability and quality.

One of the most pressing issues is the increased risk of flooding. Warmer temperatures cause polar ice to melt and sea levels to rise, putting coastal areas at greater risk of inundation. Intense rainfall and storms can overwhelm drainage systems, contaminating freshwater supplies with sewage, chemicals, and debris. The aftermath of floods often includes outbreaks of waterborne diseases due to contaminated drinking water.

On the other hand, climate change is also intensifying droughts. As heatwaves become more frequent and prolonged, rivers and reservoirs dry up, reducing access to safe drinking water. Lower water levels can concentrate pollutants, making the remaining water less safe for consumption. Additionally, droughts negatively affect agriculture, leading to food shortages and increased competition for water resources.

Another consequence of climate change is the disruption of natural water purification processes. As ecosystems are stressed, wetlands and forests, which naturally filter water, become less effective. This can lead to the spread of contaminants and pathogens, increasing the risk of illness.

Addressing climate change is essential to ensure water safety. Reducing carbon emissions, investing in resilient infrastructure, and protecting natural water sources are crucial steps in mitigating the risks and safeguarding water for future generations.

Weather events: flooding and drought

Extreme weather events, such as floods and droughts, can have devastating effects. Flooding occurs when water

overflows onto normally dry land, often due to heavy rainfall, storms, or melting snow. While floods can happen almost anywhere, areas near rivers, lakes, or coastal regions are more prone.

Floodwaters are hazardous because they can carry debris, sewage, and chemicals, contaminating drinking water sources. This can lead to outbreaks of diseases like cholera, dysentery, and hepatitis A. Fast-moving floodwaters can also sweep away vehicles and cause injuries or fatalities. Additionally, the aftermath of a flood often leaves behind damaged buildings and disrupted services, making it challenging to access clean water, food, and medical care.

Drought, on the other hand, is a prolonged period of abnormally low rainfall, resulting in a shortage of water. When water sources dry up, communities may struggle to access enough water for drinking, farming, and hygiene. This lack of water can lead to food shortages and increased risk of malnutrition. Moreover, drought conditions can reduce the quality of available water, as lower levels may concentrate pollutants and pathogens, making waterborne diseases more likely.

Contaminated water: a hidden threat

Water contamination is a significant problem globally, affecting both developed and developing regions. Water can become contaminated in various ways – through industrial discharge, agricultural runoff, sewage leaks, or natural contamination from the ground. When contaminated water is used for drinking, cooking, or washing, it poses a serious health risk.

One of the most common contaminants is bacteria, such as Escherichia coli (E. coli), which can cause severe stomach cramps, vomiting, and diarrhoea. Other harmful microorganisms include viruses like norovirus, which can spread rapidly through polluted water sources. In some cases, water may also contain toxic chemicals such as lead or pesticides, leading to long-term health problems, including developmental issues and organ damage.

Even clear and seemingly clean water can harbour invisible dangers. Microscopic parasites, like Giardia and Cryptosporidium, are common in untreated water and can cause gastrointestinal illness. Drinking or swimming in contaminated water can introduce these pathogens into the body, leading to infections that might require medical treatment.

Waterborne diseases: invisible killers

Waterborne diseases result from drinking or coming into contact with contaminated water. These diseases are often caused by bacteria, viruses, or parasites, and they can spread quickly, particularly in areas with poor sanitation or after natural disasters.

One of the most notorious waterborne diseases is cholera, caused by the bacterium Vibrio cholerae. It spreads through contaminated drinking water and can cause severe dehydration due to watery diarrhoea. Without prompt treatment, cholera can be fatal. Outbreaks are most common in areas with inadequate water treatment facilities and overcrowded living conditions.

Another disease to be aware of is hepatitis A, a viral infection that affects the liver. It is commonly transmitted through the ingestion of water contaminated with faecal matter. Symptoms include fever, nausea, abdominal pain, and jaundice (yellowing of the skin and eyes). Although hepatitis A usually resolves on its own, severe cases may require medical intervention.

Parasitic infections, like those caused by Schistosoma worms, can occur when people wade or swim in contaminated freshwaters, such as lakes or rivers. The larvae penetrate the skin and grow into adult worms within the body, leading to symptoms such as fever, abdominal pain, and organ damage over time.

Staying safe: practical precautions

To reduce the risks associated with water, it is important to follow safety guidelines and maintain good hygiene practices. Here are some practical tips to help protect against water-related dangers:

- **Avoid floodwaters:** Do not walk, swim, or drive through flooded areas. Even shallow water can be fast-moving and potentially life-threatening.
- Boil or filter water: In situations where the water source is questionable, boiling it for at least one minute can kill most bacteria, viruses, and parasites. Alternatively, use certified water purification tablets or filters.
- **Check for contamination warnings:** Local health departments often issue alerts if water quality has been compromised. Follow guidance on whether to boil water or avoid using it altogether.
- **Practice good hygiene:** Wash hands thoroughly with soap and clean water, especially after using the toilet or before preparing food. This helps prevent the spread of waterborne diseases.
- **Be cautious when swimming:** Avoid swimming in lakes, rivers, or poorly maintained pools, particularly if you have open cuts or wounds. Always shower after swimming to reduce the risk of infection.

The importance of clean water access

Access to clean, safe water is essential for good health. Yet, in many parts of the world, water scarcity and contamination remain significant challenges. Addressing these issues requires a collective effort, including better infrastructure, sustainable water management, and public awareness.

Understanding the dangers of water and how to protect oneself is crucial, especially in areas prone to natural disasters or where water treatment systems are lacking. By staying informed and taking proactive steps, it is possible to reduce the risks and maintain health and wellbeing.

Drinking water

Key facts

- In 2021, over 2 billion people live in water-stressed countries, which is expected to be exacerbated in some regions as result of climate change and population growth (1).
- In 2022, globally, at least 1.7 billion people use a drinking water source contaminated with faeces. Microbial contamination of drinking water as a result of contamination with faeces poses the greatest risk to drinking water safety.
- Safe and sufficient water facilitates the practice of hygiene, which is a key measure to prevent not only diarrhoeal diseases, but acute respiratory infections and numerous neglected tropical diseases.
- Microbiologically contaminated drinking water can transmit diseases such as diarrhoea, cholera, dysentery, typhoid and polio and is estimated to cause approximately 505 000 diarrhoeal deaths each year.
- In 2022, 73% of the global population (6 billion people) used a safely managed drinking water service – that is, one located on premises, available when needed, and free from contamination.

Overview

Safe and readily available water is important for public health, whether it is used for drinking, domestic use, food production or recreational purposes. Improved water supply and sanitation, and better management of water resources, can boost countries' economic growth and can contribute greatly to poverty reduction.

In 2010, the UN General Assembly explicitly recognised the human right to water and sanitation. Everyone has the right to sufficient, continuous, safe, acceptable, physically accessible and affordable water for personal and domestic use.

Drinking water services

Sustainable Development Goal target 6.1 calls for universal and equitable access to safe and affordable drinking water. The target is tracked with the indicator of 'safely managed drinking water services' – drinking water from an improved water source that is located on premises, available when needed, and free from faecal and priority chemical contamination.

In 2022, 6 billion people used safely managed drinking water services – that is, they used improved water sources located on premises, available when needed, and free from contamination. The remaining 2.2 billion people without safely managed services in 2022 included:

- 1.5 billion people with basic services, meaning an improved water source located within a round trip of 30 minutes;
- 292 million people with limited services, or an improved water source requiring more than 30 minutes to collect water;
- 296 million people taking water from unprotected wells and springs; and
- 115 million people collecting untreated surface water from lakes, ponds, rivers and streams.

Sharp geographic, sociocultural and economic inequalities persist, not only between rural and urban areas but also in towns and cities where people living in low-income, informal or illegal settlements usually have less access to improved sources of drinking water than other residents.

Water and health

Contaminated water and poor sanitation are linked to transmission of diseases such as cholera, diarrhoea, dysentery, hepatitis A, typhoid and polio. Absent, inadequate, or inappropriately managed water and sanitation services expose individuals to preventable health risks. This is particularly the case in health care facilities where both patients and staff are placed at additional risk of infection and disease when water, sanitation and hygiene services are lacking.

Out of every 100 patients in acute-care hospitals, 7 patients in high-income countries (HICs) and 15 patients in low- and middle-income countries (LMICs) will acquire at least one health care-associated infection during their hospital stay.

Inadequate management of urban, industrial and agricultural wastewater means the drinking water of hundreds of millions of people is dangerously contaminated or chemically polluted. Natural presence of chemicals, particularly in groundwater, can also be of health significance, including

arsenic and fluoride, while other chemicals, such as lead, may be elevated in drinking water as a result of leaching from water supply components in contact with drinking water.

Some 1 million people are estimated to die each year from diarrhoea as a result of unsafe drinking water, sanitation and hand hygiene. Yet diarrhoea is largely preventable, and the deaths of 395,000 children aged under 5 years could be avoided each year if these risk factors were addressed. Where water is not readily available, people may decide handwashing is not a priority, thereby adding to the likelihood of diarrhoea and other diseases.

Diarrhoea is the most widely known disease linked to contaminated food and water but there are other hazards. In 2021, over 251.4 million people required preventative treatment for schistosomiasis – an acute and chronic disease caused by parasitic worms contracted through exposure to infested water.

In many parts of the world, insects that live or breed in water carry and transmit diseases such as dengue fever. Some of these insects, known as vectors, breed in clean, rather than dirty water, and household drinking water containers can serve as breeding grounds. The simple intervention of covering water storage containers can reduce vector breeding and may also reduce faecal contamination of water at the household level.

Economic and social effects

When water comes from improved and more accessible sources, people spend less time and effort physically collecting it, meaning they can be productive in other ways. This can also result in greater personal safety and reducing musculoskeletal disorders by reducing the need to make long or risky journeys to collect and carry water. Better water sources also mean less expenditure on health, as people are less likely to fall ill and incur medical costs and are better able to remain economically productive.

With children particularly at risk from water-related diseases, access to improved sources of water can result in better health, and therefore better school attendance, with positive longer-term consequences for their lives.

Challenges

Historical rates of progress would need to double for the world to achieve universal coverage with basic drinking water services by 2030. To achieve universal safely managed services will require a six-fold increase. Climate change, increasing water scarcity, population growth, demographic changes and urbanisation already pose challenges for water supply systems. Over 2 billion people live in water-stressed countries, which is expected to be exacerbated in some regions as result of climate change and population growth. Re-use of wastewater to recover water, nutrients or energy is becoming an important strategy. Use of wastewater and sludge is widespread globally; however, much is used informally and/or without sufficient treatment and other controls to ensure that human and environmental health is protected. If done appropriately, safe use of wastewater and sludge can yield multiple benefits, including increased food production, increased resilience to water and nutrient scarcity and greater circularity in the economy.

Options for water sources used for drinking water and irrigation will continue to evolve, with an increasing reliance on groundwater and alternative sources, including wastewater. Climate change will lead to greater fluctuations in harvested rainwater. Management of all water resources will need to be improved to ensure provision and quality.

13 September 2023

www.who.int

What contaminants lurk in the UK's drinking water? An expert explains

An article from The Conversation.

By Vanessa Speight, Professor of Integrated Water Systems, University of Sheffield

Recently, a school project made an alarming discovery: the presence of five times the recommended maximum amount of lead in water samples taken from 14 schools across the UK. Lead is a toxin which even at low levels is capable of affecting children's brain development and reducing their IQ. The news might well make the British public worry about what exactly is lurking in their drinking water.

Installing lead pipes in the UK's drinking water network has been banned for decades, but about eight million old buildings may still have lead pipes in service. Usually, external lead pipes connect the building to the main water pipe, but in some cases, internal plumbing can also contain lead pipes.

Brass plumbing fittings, such as those used in water fountains, can contain low levels of lead and can release that lead if water sits in them for a long time – or if chemicals like chloride make the water corrosive, meaning bits of pipe start leaking into the water. To help prevent this, water companies in the UK add corrosion inhibitors, such as orthophosphate, to drinking water.

Meanwhile, UK water companies are working to identify and remove lead pipes across the country – but this work is slow and expensive. Fundamentally, while lead pipes are in service, there remains a risk of contamination.

Building owners should take action to protect vulnerable people, particularly children, by checking for the presence of lead pipes and testing their water. This can be done by private laboratories, but many water companies will also perform this test if asked to do so.

The US has announced a high-profile infrastructure renewal plan that commits to removing all lead pipes as part of a £39.7 billion water system upgrade fund over the next five years: an example the UK government should consider following.

But lead is not the only contaminant that might be leaking into people's homes. An increasingly polluted environment means that many contaminants make their way into rivers, lakes and, ultimately, drinking water.

Emerging contaminants

One example of contaminants that have emerged as a health concern in drinking water is a group of chemicals collectively referred to as perfluoroalkyl and polyfluoroalkyl substances (PFAS). Historically used as nonstick coatings for items such as cooking pans and in fire-fighting foams, PFAS can now be found wherever humans live.

PFAS are sometimes called 'forever chemicals' because they don't naturally degrade in the environment, instead they accumulate in the bodies of organisms like fish.

There is considerable scientific debate around the world about how many PFAS can be present in drinking water and food before they pose a danger to human health.

In England, the Drinking Water Inspectorate has set guidance levels for two PFAS compounds: perfluorooctane sulphonate (PFOS) and perfluorooctanoic acid (PFOA). If either compound is detected above 0.01 micrograms per litre of water, water companies must perform a risk assessment, test all their drinking water supplies, and report their findings to local health authorities.

Keeping up with contaminants

The water industry is continually trying to keep up with the latest potential contaminants – including microbiological contaminants including diarrhoea-causing parasites such as cryptosporidium, and ever-increasing volumes of microplastics – to understand where they come from, how to remove them, and what they do inside the human body.

It takes years to study a new contaminant: from determining how best to detect it, understanding how much of it is present in water compared with other sources such as food, and quantifying its effect on public health to justify spending money on regulation.

Drinking water treatment continues to improve, but many of the newer technologies needed to remove contaminants like PFAS require even more energy and chemicals than current methods use. From a sustainability perspective, the best course of action is to prevent these contaminants from entering drinking water sources in the first place.

In the UK, many contaminants are both legally and illegally dumped in waterways in much higher concentrations than are allowed for drinking water. Regulations should be coordinated to make sure that unsafe levels of contaminants don't get anywhere near the water systems.

Historically, the approach to pollution was to rely on dilution in rivers and lakes to reduce concentrations of potentially dangerous substances. Now, as we understand more about the potential environmental and human health effects of the toxins around us, it's time to introduce more stringent controls.

14 September 2021

Seven common waterborne diseases (and how to prevent them)

Each year, waterborne diseases afflict hundreds of millions of people, primarily those living without safe, accessible water in developing countries.

Of the seven most common waterborne diseases in the world, diarrhoea is the central symptom. The latest research shows that diarrhoea is the second leading cause of death for children under the age of five, causing more childhood deaths than malaria, AIDS, and measles combined.

That's hundreds of thousands of deaths, but there is hope for the future. Experts believe we can end the global water and sanitation crisis in our lifetime.

What are waterborne diseases?

Waterborne diseases are illnesses caused by microscopic organisms, like viruses and bacteria, that are ingested through contaminated water or by coming in contact with faeces.

If every person on the planet was able to practice safe sanitation and hygiene and have access to clean water, these diseases would not exist. Governments, NGOs, and communities themselves have made great strides in the past 20 years to end waterborne diseases. Still, there is much to be done.

Learn about seven waterborne diseases and help prevent them today.

1. Typhoid Fever

Although rare in industrialised countries, typhoid fever is well-known in extremely poor parts of developing nations; it's estimated that up to 20 million people worldwide suffer from the illness each year. It's spread through contaminated food, unsafe water, and poor sanitation, and it is highly contagious.

Symptoms include:

- A fever that increases gradually
- Muscle aches
- Fatigue
- Sweating
- Diarrhoea or constipation

Prevention and treatment

Vaccines are recommended for people who are travelling in areas where poor sanitation and unsafe water are common. The vaccine can be injected via a shot or taken orally for a number of days. To prevent it, refrain from drinking any water that isn't bottled and sealed, and do not eat food from villages or street vendors. Typhoid is treated with antibiotics.

2. Cholera

Cholera is commonly found in humanitarian emergencies or marginalised villages where poverty and poor sanitation are rampant. The disease is spread through contaminated water and causes severe dehydration and diarrhoea. Cholera can be fatal within days or even hours of exposure to the bacteria, but only 1 in 10 people will develop life-threatening symptoms.

Symptoms include:

- Nausea
- Vomiting
- Diarrhoea
- Muscle cramps

Prevention and treatment

Cholera is a waterborne illness that's easily prevented when travelling. Wash your hands often, only eat foods that are completely cooked and hot (no sushi), and only eat vegetables you can peel yourself, like avocados, bananas, and oranges. Of course, drink safe water.

When handwashing in unavailable, cholera can impact an entire village. In developing countries like Ethiopia, data shows that 40% of households do not have means to wash their hands properly, meaning they don't have safe water, soap, and a facility to wash. This makes hygiene management and disease prevention nearly impossible for these communities.

Water For Good helps prevent cholera in remote villages by teaching families how to construct their own handwashing devices. To date, 5,970 homes in Ethiopia alone have built their own handwashing station (called a 'tippy tap') using locally-sourced materials.

3. Giardia

This waterborne disease is shared through contaminated water, most often in ponds and streams, but it can also be found in a town's water supply, swimming pools, and more. The infection is caused by a parasite and typically clears up after a few weeks. However, it's possible those who have been exposed will experience intestinal problems for years to come.

Symptoms include:

- Abdominal pain
- Cramps and bloating
- Diarrhoea
- Nausea
- Weight loss

Prevention and treatment

While there is no vaccine for giardia, there are simple ways to avoid the infection. Wash your hands with soap often, don't swallow water while swimming, and drink only bottled water.

With time, the immune system will typically beat giardia on its own. But, if symptoms worsen, doctors prescribe anti-parasite and antibiotic medications.

Water-poor communities cannot protect themselves from illnesses like giardia, and treatment for this illness can come at a high cost for a family living in poverty. For these reasons, Water For Good's programs focus on long-term prevention. This includes constructing safe water sources and teaching health practices, one house at a time, until the entire community has the resources and the knowledge to prevent waterborne illness.

When families learn how to construct their own handwashing facilities, bathrooms, and dish drying racks, they take control of their health. They check off a list of basic health practices, and they become certified Water For Good 'Healthy Homes.'

4. Dysentery

An intestinal infection, dysentery is a waterborne disease characterised by severe diarrhoea as well as blood or mucus in the stool. Dysentery is a good reason to always wash your hands, as the disease is spread mainly through poor hygiene. It can be caused by bacteria, viruses, or parasites in unsafe food and water and by people coming in contact with faecal matter. If someone experiencing dysentery cannot replace fluids quickly enough, their life could be at risk.

Symptoms include:

- Stomach cramps and pain
- Diarrhoea

- Fever
- Nausea
- Vomiting
- Dehydration

Prevention and treatment

To prevent dysentery, wash your hands with soap frequently, order all drinks without ice, don't eat food sold by street vendors, and only eat fruits you can peel. Drink only sealed, bottled water while travelling in places with higher dysentery risk, such as communities where proper hygiene practices are uncommon.

Mild dysentery usually clears up with rest and fluids, but over-the-counter medications such as Pepto-Bismol can help with stomach cramping. More severe cases can be treated with antibiotics, although some strains of the disease are resistant.

5. Escherichia Coli (E. coli)

E. coli is a bacteria with various strains, some dangerous and some beneficial. For example, E. coli bacteria is important in creating a healthy intestinal tract.

However, if animal waste has found its way into farmland where produce is grown or if strains of E. coli are spread through the process of making ground beef, those who consume these foods could experience symptoms of the waterborne illness. The bacteria is also found in unsafe water sources around the globe where human water sources and cattle coexist.

Symptoms of dangerous strains of E. coli are similar to that of dysentery and other waterborne diseases. Most bouts of E. coli pass within a week, but older people and young children have a greater chance of developing life-threatening symptoms. Anyone believed to have been exposed to contaminated food or water should contact a doctor if diarrhoea contains blood.

Prevention and treatment

As always, avoid water possibly contaminated by human and/or animal faeces (like ponds, rivers, and swamps). If you are going to eat ground beef, cook thoroughly. Wash fruits and vegetables well, wash hands often, and drink only safe water.

To treat the disease, drink plenty of safe water, rest, and take over-the-counter diarrheal medication.

While these are simple prevention and treatment tips, there are many remote communities in Uganda who have no choice but to drink from swamps.

Water For Good staff are serving the village of Kikomera Biri, Uganda, where families gather water from a swamp. The results of water testing showed an extremely high risk for dangerous pathogens like typhoid, harmful strains of E. coli, and other waterborne diseases. Unless this community – which is already experiencing extreme poverty – pays for a taxi to drive into town for expensive, bottled water, they have no choice but to keep drinking from the swamp.

Thankfully, a new safe water source for all 299 residents is planned for construction this year.

6. Hepatitis A

Hepatitis A is a liver infection caused by consuming contaminated food and water or by coming in close contact with someone who has the infection. People who travel in developing countries often or work in rural communities with poor sanitation and hygiene management are most exposed to the disease.

Symptoms include:

- Fatigue
- Clay-coloured bowel movements
- Jaundice
- Nausea and vomiting
- Abdominal pain, especially near your liver
- Loss of appetite
- Sudden fever

The infection usually goes away in a few weeks, but it's possible that it can become severe and last several months.

Prevention and treatment

The best way to prevent hepatitis A is by getting the vaccine. Eat only foods that are thoroughly cooked and served hot, and avoid eating anything at room temperature. Only eat fruit that you can peel and that you have peeled yourself. Don't eat from food vendors and don't eat runny eggs or raw/rare meat.

Once a person has hepatitis A, they build an immunity and will likely never get it again. However, the symptoms are serious, often forcing people to take time off work or school to recover. If you have contracted hepatitis A, rest, avoid drinking alcohol, and drink plenty of fluids. The disease will run its course, and full recovery is expected after three months.

7. Salmonella

Most cases of salmonella come from ingesting food or water contaminated with faeces. Undercooked meat, egg products, fruits, and vegetables can also carry the disease. Most people don't develop complications, but children, pregnant women, older adults, and people with weakened immune systems are most at risk.

Symptoms include:

- Blood in stool
- Chills
- Headache
- Diarrhoea

Prevention and treatment

When preparing your own food, make sure to cook thoroughly and store or freeze within 30 minutes of use. Avoid touching birds or reptiles, and as always, wash your hands frequently.

Salmonella infection dehydrates the body. Treat it by drinking fluids and electrolytes. More serious infections can require hospitalisation and antibiotics.

Help prevent waterborne diseases: support Water For Good's work

There are many parts in the world where waterborne diseases are rampant, deadly, and knowledge about prevention is not widely available. For over 40 years, Water For Good has sought out these places, working with communities to teach vital sanitation and health practices and constructing custom water technologies in places where water access is most difficult.

Over and over again, cholera is prevented and typhoid eradicated. Children no longer battle waterborne illness, and parents go back to work.

The above information is reprinted with kind permission from Water For Good.

www.waterforgood.org

How climate change affects waterborne diseases

Waterborne diseases like cholera and typhoid are still a leading cause of human morbidity and mortality worldwide. As climate change increasingly impacts global temperatures and weather events, the risk of these infectious diseases will only worsen. But there are actions we can take to reduce these risks.

Summary

What are the effects of climate change on waterborne diseases?

- Waterborne diseases cause 3.5 million deaths annually.
- High temperatures, heavy rain and drought increase the risk of waterborne diseases, such as cholera, typhoid and E. coli.
- Actions to mitigate and adapt to climate change, such as transitioning to renewable energy and improving disease surveillance, can help reduce the risk of waterborne diseases.

What causes waterborne diseases?

Waterborne diseases are caused by drinking or coming into contact with contaminated water.

This contamination could be caused by pollutants, like agricultural chemicals, or faeces from an infected person or animal.

How is climate change increasing the risk of waterborne diseases?

Waterborne diseases, such as cholera, typhoid and E. coli infections, are a major public health issue causing 3.5 million deaths annually. And climate change is increasing the risk.

The climate variables that most influence waterborne illness include:

- extreme weather events (like storms, droughts, precipitation, floods and wildfires)
- increase in air temperature
- increase in water temperature

What are the risks?

Cholera outbreaks are closely related to environmental conditions. Cholera is an acute diarrhoeal infection caused by ingestion of food or water contaminated with the bacterium Vibrio cholerae.

Temperature and precipitation can be influential in cholera outbreaks. High temperature can accelerate the growth and proliferation of pathogens in their habitats (like raw food or water). Hotter temperatures can also increase the risk of droughts which can lead to poor sanitation and cholera outbreaks. For example, prolonged and extreme drought in Kenya triggered a cholera outbreak in 2022. More than 7,800 cholera cases and 122 deaths were reported by March 2023.

Rainfall also has a direct influence on the transmission of cholera. High rainfall can increase the risk of wastewater contaminating either raw or treated water (person-to-environment transmission). While low rainfall can increase the concentration of pathogens in water (environment-to-person transmission).

Extreme weather events can also result in the ideal conditions for these infectious diseases to spread. In 2022, Malawi was hit by two tropical storms, Ana and Gombe, during a heavy rainy season. Homes were destroyed and nearly 200,000 people were displaced. This combination of events sparked a cholera outbreak. The disease, which is usually endemic in Malawi, continued to spread through the dry season. By February 2023, 36,943 cases and 1,210 deaths had been reported, marking the biggest outbreak in the country's history.

Cholera isn't the only waterborne disease impacted by high temperature, heavy rain and drought. For example, one study in Israel found that temperature increases the incidence of E. coli bloodstream infections. And in Fiji, cases of typhoid are more likely to occur in areas that have experienced flooding of a stream or river.

What can be done to reduce the risk?

Climate change is increasing the risk of waterborne diseases. The sooner we act to mitigate the impacts of climate change – by transitioning from using fossil fuels to clean, renewable energy – the better off we'll be in the future.

But mitigation alone will not be enough. There are other actions we must take to adapt to the risks:

- provide universal access to care and disease management
- improve disease surveillance
- develop and invest in early warning systems to monitor changes in climatic conditions
- upgrade water catchment, storage, treatment and distribution systems
- protect critical infrastructure from floods, storms and sea level rise
- limit water overuse
- use household water purification systems

12 May 2022

Global water crisis leaves half of world food production at risk in next 25 years

Landmark review says urgent action needed to conserve resources and save ecosystems that supply fresh water.

By Fiona Harvey, Environment editor

More than half the world's food production will be at risk of failure within the next 25 years as a rapidly accelerating water crisis grips the planet, unless urgent action is taken to conserve water resources and end the destruction of the ecosystems on which our fresh water depends, experts have warned in a landmark review.

Half the world's population already faces water scarcity, and that number is set to rise as the climate crisis worsens, according to a report from the Global Commission on the Economics of Water published on Thursday.

Demand for fresh water will outstrip supply by 40% by the end of the decade, because the world's water systems are being put under 'unprecedented stress', the report found.

The commission found that governments and experts have vastly underestimated the amount of water needed for people to have decent lives. While 50 to 100 litres a day are required for each person's health and hygiene, in fact people require about 4,000 litres a day in order to have adequate nutrition and a dignified life. For most regions, that volume cannot be achieved locally, so people are dependent on trade – in food, clothing and consumer goods – to meet their needs.

Some countries benefit more than others from 'green water', which is soil moisture that is necessary for food production, as opposed to 'blue water' from rivers and lakes. The report found that water moves around the world in 'atmospheric rivers' which transport moisture from one region to another.

About half the world's rainfall over land comes from healthy vegetation in ecosystems that transpires water back into the atmosphere and generates clouds that then move downwind. China and Russia are the main beneficiaries of these 'atmospheric river' systems, while India and Brazil are the major exporters, as their landmass supports the flow of green water to other regions. Between 40% and 60% of the source of fresh water rainfall is generated from neighbouring land use.

'The Chinese economy depends on sustainable forest management in Ukraine, Kazakhstan and the Baltic region,' said Prof Johan Rockström, the director of the Potsdam Institute for Climate Impact Research and one of the co-chairs of the commission. 'You can make the same case for Brazil supplying fresh water to Argentina. This interconnectedness just shows that we have to place fresh water in the global economy as a global common good.'

Tharman Shanmugaratnam, the president of Singapore and a co-chair of the commission, said countries must start cooperating on the management of water resources before it was too late.

'We have to think radically about how we are going to preserve the sources of fresh water, how we are going to use it far more efficiently, and how we are going to be able to have access to fresh water available to every community, including the vulnerable – in other words, how we preserve equity [between rich and poor],' Shanmugaratnam said.

The Global Commission on the Economics of Water was set up by the Netherlands in 2022, drawing on the work of dozens of leading scientists and economists, to form a comprehensive view of the state of global hydrological systems and how they are managed. Its 194-page report is the biggest global study to examine all aspects of the water crisis and suggest remedies for policymakers.

The findings were surprisingly stark, said Rockström. 'Water is victim number one of the [climate crisis], the environmental changes we see now aggregating at the global level, putting the entire stability of Earth's systems at risk,' he told the Guardian. '[The climate crisis] manifests itself first and foremost in droughts and floods. When you think of heatwaves and fires, the really hard impacts are via moisture – in the case of fires, [global heating] first dries out landscapes so that they burn.'

Every 1C increase in global temperatures adds another 7% of moisture to the atmosphere, which has the effect of 'powering up' the hydrological cycle far more than would happen under normal variations. The destruction of nature is also further fuelling the crisis, because cutting down forests and draining wetlands disrupts the hydrological cycle that depends on transpiration from trees and the storage of water in soils.

Harmful subsidies are also distorting the world's water systems, and must be addressed as a priority, the experts found. More than $700 billion (£540 billion) of subsidies each year go to agriculture, and a high proportion of these are misdirected, encouraging farmers to use more water than they need for irrigation or in wasteful practices.

Industry also benefits – about 80% of the wastewater used by industries around the world is not recycled.

Ngozi Okonjo-Iweala, the director general of the World Trade Organization, also a co-chair of the commission, said countries must redirect the subsidies, axing harmful ones while ensuring poor people were not disadvantaged. 'We must have a basket of policy tools working together if we are to get the three Es – efficiency, equity and environmental sustainability and justice. Therefore we have to couple the pricing of water with appropriate subsidies,' she said.

At present, subsidies mainly benefit those who are better off, Okonjo-Iweala added. 'Industry is getting a lot of the subsidy, and richer people. So what we need are better targeted subsidies. We need to identify the poor people who really need this,' she said.

Developing countries must also be given access to the finance they need to overhaul their water systems, provide safe water and sanitation, and halt the destruction of the natural environment, the report found.

Mariana Mazzucato, professor of economics at University College London, and a co-chair of the commission, said loans made by public sector banks to developing countries should be made conditional on water reforms. 'These could be improving water conservation and the efficiency of water use, or direct investment for water-intensive industries,' she said. '[We must ensure] profits are reinvested in productive activity such as research and development around water issues.'

Water problems also had an outsized impact on women and girls, Mazzucato added. 'One of our commissioners is Yvonne Aki-Sawyerr, the mayor of Freetown in Sierra Leone. She says most of the rapes and abuse of women actually happen when they're going to fetch water,' Mazzucato said. 'Child mortality, gender parity, the water collection burden, the food security burden – they're all connected.'

Five main takeaways from the report

The world has a water crisis

More than 2 billion people lack access to safe drinking water, and 3.6 billion people – 44% of the population – lack access to safe sanitation. Every day, 1,000 children die from lack of access to safe water. Demand for fresh water is expected to outstrip its supply by 40% by the end of this decade. This crisis is worsening – without action, by 2050 water problems will shave about 8% off global GDP, with poor countries facing a 15% loss. Over half of the world's food production comes from areas experiencing unstable trends in water availability.

There is no coordinated global effort to address this crisis

Despite the interconnectedness of global water systems there are no global governance structures for water. The UN has held only one water conference in the past 50 years, and only last month appointed a special envoy for water.

Climate breakdown is intensifying water scarcity

The impacts of the climate crisis are felt first on the world's hydrological systems, and in some regions those systems are facing severe disruption or even collapse. Drought in the Amazon, floods across Europe and Asia, and glacier melt in mountains, which causes both flooding and droughts downstream, are all examples of the impacts of extreme weather that are likely to get worse in the near future. People's overuse of water is also worsening the climate crisis – for instance, by draining carbon-rich peatlands and wetlands that then release carbon dioxide into the atmosphere.

Water is artificially cheap for some and too expensive for others

Subsidies to agriculture around the world often have unintended consequences for water, providing perverse incentives for farmers to over-irrigate their crops or use water wastefully. Industries also have their water use subsidised, or their pollution ignored, in many countries. Meanwhile, poor people in developing countries frequently pay a high price for water, or can only access dirty sources. Realistic pricing for water that removes harmful subsidies but protects the poor must be a priority for governments.

Water is a common good

All of human life depends on water, but it is not recognised for the indispensable resource it is. The authors of the report urge a rethink of how water is regarded – not as an endlessly renewable resource, but as a global common good, with a global water pact by governments to ensure they protect water sources and create a 'circular economy' for water in which it is reused and pollution cleaned up. Developing nations must be given access to finance to help them end the destruction of natural ecosystems that are a key part of the hydrological cycle.

16 October 2024

Glacier meltdown risks food and water supply of 2 billion people, says UN

Unesco report highlights 'unprecedented' glacier loss driven by climate crisis, threatening ecosystems, agriculture and water sources.

By Fiona Harve,y Environment editor

Retreating glaciers threaten the food and water supply of 2 billion people around the world, the UN has warned, as current 'unprecedented' rates of melting will have unpredictable consequences.

Two-thirds of all irrigated agriculture in the world is likely to be affected in some way by receding glaciers and dwindling snowfall in mountain regions, driven by the climate crisis, according to a Unesco report.

More than 1 billion people live in mountainous regions and, of those in developing countries, up to half are already experiencing food insecurity. That is likely to worsen, as food production in such regions is dependent on mountain waters, melting snow and glaciers, according to the World Water Development Report 2025.

Developed countries are also at risk: in the US, for example, the Colorado River basin has been in drought since 2000, and higher temperatures mean more of the precipitation is falling as rain, which runs off more quickly than mountain snow, exacerbating drought conditions.

Audrey Azoulay, the director general of Unesco, said: 'Regardless of where we live, we all depend in some way on mountains and glaciers. But these natural water towers are facing imminent peril. This report demonstrates the urgent need for action.'

The rate of change of glaciers is the worst on record, according to separate research from the World Meteorological Organisation, which published its annual State of the Climate report this week. The largest three-year loss of glacier mass on record occurred in the past three years, the study found, with Norway, Sweden, Svalbard and the tropical Andes among the worst-affected areas.

Eastern Africa has lost 80% of its glaciers in places and, in the Andes, between a third and a half of glaciers have melted since 1998. Glaciers in the Alps and the Pyrenees, the worst affected in Europe, have shrunk by about 40% over roughly the same period.

The decline of glaciers has had a further impact, added Abou Amani, director of water sciences at Unesco, in that the loss of ice replaces a reflective surface with dark soil that absorbs heat. 'Glaciers melting have an impact on the reflectivity of [solar] radiation and that will impact the whole climate system,' he warned.

Icebergs manifest in seven distinct types, categorised as detached from glaciers, floating, or grounded. These types exhibit diverse characteristics, including pointed tops, domes, erosion patterns, pooling, inclination, plate-like formations, and block shapes.

More avalanches will also result, as rain falling on snow is a major factor behind avalanche formation. Pooling water from melting glaciers can also be released, causing sudden floods in valleys or to people living further down the slopes. Permafrost is also melting, releasing methane from the mountain soils that melting glaciers are uncovering.

A previous study, published last month in the peer-reviewed journal Nature, found that half of global glacier mass would be lost by the end of the century, if global heating was not halted. Alex Brisbourne, a glacier geophysicist at the British Antarctic Survey, said: 'Mountain glaciers contain some of the largest freshwater reservoirs on Earth. Meltwater released in the summer provides the water supply to a billion people and sustains an enormous amount of industry and agriculture. The impact [of such melting] will be felt way beyond those immediately downstream of the glaciers.'

These impacts are coming at a time when many food sources are already under strain. Alvaro Lario, president of the International Fund for Agricultural Development (Ifad) and chair of UN-Water, called for more support for people who live in afflicted mountain regions. 'Water flows downhill, but food insecurity rises uphill. Mountains provide 60% of our freshwater, but the communities that safeguard these vital resources are among the most food insecure,' he said.

'We must invest in their resilience to protect glaciers, rivers, and a shared future for all of us.'

20 March 2025

From deluges to drought: climate change speeds up water cycle, triggers more extreme weather

Around the globe, hotter temperatures stoked by climate change are increasing the odds of both severe drought and heavier precipitation that can wreak havoc on people and the environment.

By Tammy Webber and Donavon Brutus

Prolonged droughts, wildfires and water shortages. Torrential downpours that overwhelm dams and cause catastrophic flooding.

Around the globe, rising temperatures stoked by climate change are increasing the odds of both severe drought and heavier precipitation that wreak havoc on people and the environment.

Rainfall can disappear for years only to return with a vengeance, as it did in California in 2023, with record-setting rain and snowfall. That led to heavy vegetation growth that provided fuel for the devastating January wildfires in Los Angeles after drought returned.

But how can global warming cause both drier and wetter extremes? Here's what experts say.

It's all about the water cycle

Water constantly moves between the Earth and its atmosphere. But that system – called the hydrological cycle – is speeding up as global temperatures get hotter, primarily due to the burning of fossil fuels like coal and gas.

A hotter atmosphere sucks up more water vapour from bodies of water and vegetation and soil.

Over land, this atmospheric demand and loss of surface moisture leads to longer and more intense droughts, even causing some arid areas to expand. Though rain falls less often, when it does, it's often in intense and destructive deluges.

That's because the atmosphere holds 7% more water vapour for every degree Celsius.

'Basically, global warming is turning the atmosphere into a bigger sponge so it can soak up more moisture ... and then when the conditions are right for rainfall, it's like squeezing that sponge,' said Jonathan Overpeck, a climate scientist at the University of Michigan. 'You get more moisture coming out faster.'

Oceans play outsized role

Oceans absorb most of the planet's extra heat. That causes the water to expand and ice to melt at the poles, raising sea levels. The warmer water also provides fuel for larger hurricanes and cyclones that can dump massive amounts of water in a short time.

In 2023, for example, heavy one-day rains from Mediterranean storm Daniel caused massive flooding across eastern Libya that overwhelmed two dams, sending a wall of water through the coastal city of Derna that destroyed entire neighbourhoods and swept bridges, cars and people out to sea. Climate scientists say climate change made that storm far more likely.

Snowpack is diminishing

Climate change also is affecting snowpack, a critical part of the hydrological cycle.

Melting snow helps fill reservoirs and waterways, including for drinking and agriculture. But less snow is falling in general, and what does often is absorbed by thirsty soil.

What's more, because winters are becoming warmer overall, the growing season is longer, meaning snowmelt also is being lost through evapotranspiration of plants. But, just like rain, climate change also can cause more intense and sometimes damaging snowstorms.

'All this stuff is related to warming, which we know with perfect confidence is almost all due to human activity,' Overpeck said. 'The good news is, we know how to stop it if we want to.'

24 March 2025

Research

Can you find any recent news stories where climate change has caused unusual or severe weather events?

www.independent.co.uk

Deadly floods and storms affected more than 400,000 people in Europe in 2024

European State of the Climate report 'lays bare' impact of fossil fuels on continent during its hottest 12 months on record.

By Ajit Niranjan

The home-wrecking storms and floods that swept Europe last year affected 413,000 people, a report has found, as fossil fuel pollution forced the continent to suffer through its hottest year on record.

Dramatic scenes of cars piled up on inundated streets and bridges being ripped away by raging torrents were seen around the continent in 2024, with 'high' floods on 30% of the European river network and 12% crossing the 'severe' flood threshold, according to the *European State of the Climate* report.

The two most destructive examples were the deluges that tore through central Europe in September and eastern Spain in October, which accounted for more than 250 of the 335 flooding deaths recorded across the continent in 2024.

Previous studies have found the disasters were made stronger and more likely because of global heating, which lets clouds pummel the ground with more rain.

Celeste Saulo, director general of the World Meteorological Organisation (WMO), said 'every additional fraction of a degree' of temperature rise mattered, but that societies must also adapt to a hotter world.

'We are making progress but need to go further and need to go faster,' she said. 'And we need to go together.'

The report, which was published on Tuesday by the EU's Copernicus climate change service and the WMO, found the numbers of days with 'strong', 'very strong' and 'extreme heat stress' were all the second-highest on record.

South-eastern Europe experienced its longest heatwave on record in July 2024, searing more than half the region for 13 days in a row, while high heat across the continent contributed to destructive wildfires that affected 42,000 people, the report found. About one-quarter of Europe's burnt area last year came from devastating wildfires in

Almost a third of European river networks flooded in 2024

Annual percentage of European river networks experiencing flooding, 1992-2024

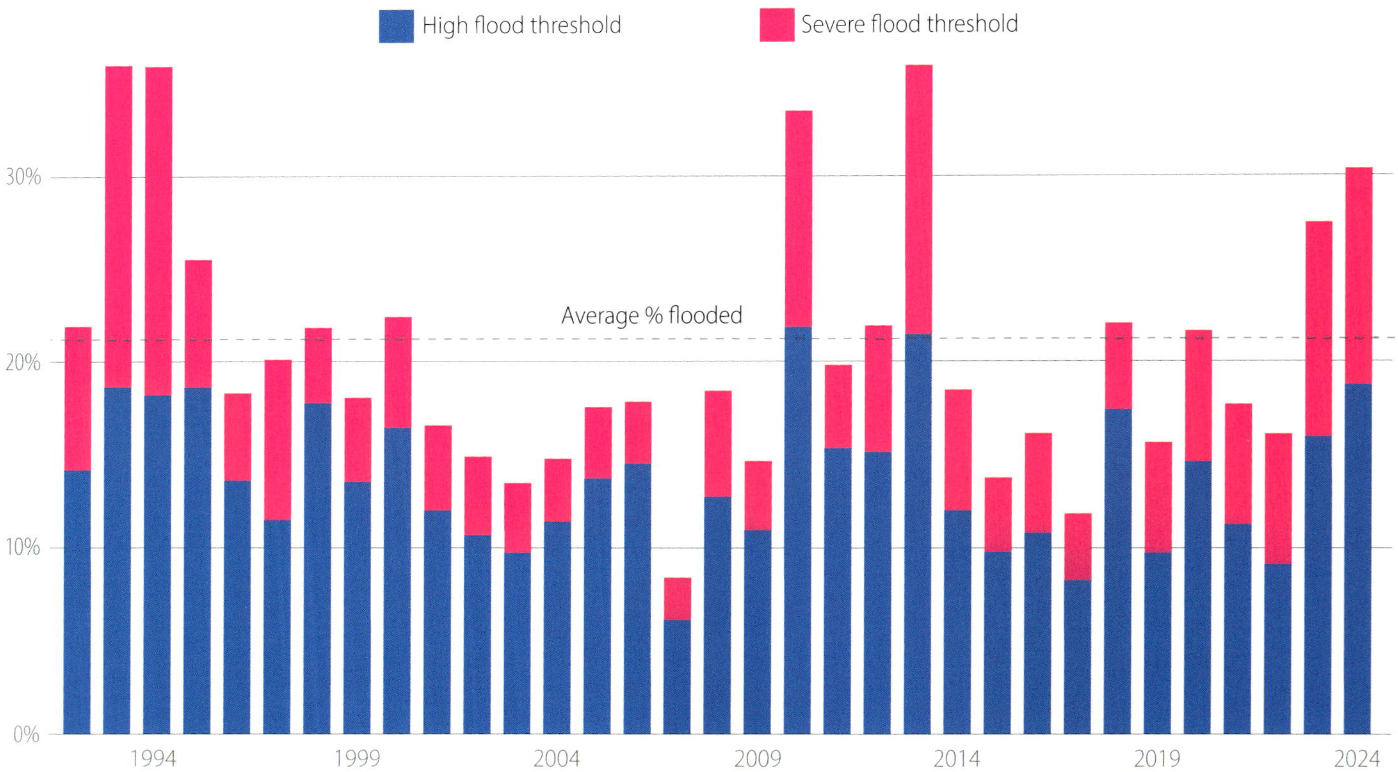

Source: EFAS, CEMS/C3S/ECMWFt

Portugal in September, which burned about 110,000 hectares in a single week.

Friederike Otto, a climate scientist at Imperial College London and co-lead of World Weather Attribution, who was not involved in the study, said the report 'lays bare the pain Europe's population is already suffering from extreme weather' at 1.3C of global heating above preindustrial levels.

'We're on track to experience 3C by 2100,' she said. 'You only need to cast your mind back to the floods in Spain, the fires in Portugal, or the summer heatwaves last year to know how devastating this level of warming would be.'

The report authors highlighted an 'unusual' contrast between western and eastern Europe, with the west tending to be wet and cloudy and the east warm and sunny. River flows tended to be above average in western countries and below average in eastern ones. In several months last year, the Thames in the UK and the Loire in France experienced their highest flows in a 33-year record, the report found.

Glaciers in all regions had net ice loss, with those in Scandinavia and Svalbard losing more mass than ever previously recorded, according to the report. The authors also noted high temperatures north of the Arctic Circle, and the hottest sea surface temperature recorded in the Mediterranean.

Froila Palmeiro, a climate scientist at the Euro-Mediterranean Center on Climate Change, who was not involved in the report, said the extremes 'not only have a direct impact on their ecosystems, but also play a role in weather patterns affecting all of Europe'.

Europe is warming twice as fast as the global average but has cut its planet-heating pollution faster than other big economies. The EU plans to hit net zero greenhouse gas emissions by 2050, and is expected to also announce a net 90% reduction target for 2040 later this year.

Thomas Gelin, a climate campaigner at Greenpeace EU, said the report showed that politicians had failed to hold fossil fuel companies accountable and stop the expansion of their polluting businesses.

'The only parts of Europe that aren't being boiled dry are being washed away in floods,' he said. 'The EU must urgently update its climate targets to reflect the scientific reality, and put a stop to new fossil fuel projects as a first step to a full phase-out.'

15 April 2025

www.theguardian.com

Drought can hit almost anywhere: How 5 cities that nearly ran dry got water use under control

An article from The Conversation.

By Sara Hughes, Adjunct Professor of Environment and Sustainability, University of Michigan and Michael Wilson, Professor of Policy Analysis, RAND School of Public Policy

Water scarcity is often viewed as an issue for the arid American West, but the US Northeast's experience in 2024 shows how severe droughts can occur in just about any part of the country.

Cities in the Northeast experienced record-breaking drought conditions in the second half of 2024 after a hot, dry summer in many areas. Wildfires broke out in several states that rarely see them.

By December, much of the region was experiencing moderate to severe drought. Residents in New York City and Boston were asked to reduce their water use, while Philadelphia faced risk to its water supply due to saltwater coming up the Delaware River.

Before the drought, many people in the region weren't prepared for water shortages or even paying much attention to their water use.

How cities have responded to drought-caused water crises

Cape Town, South Africa; Melbourne, Australia; Las Vegas, São Paulo and New Orleans have all seen severe water shortages. Here are some of the ways they are avoiding a repeat of the past.

	Cape Town	Las Vegas	São Paulo	New Orleans	Melbourne
Infrastructure					
Repair leaks	■	■	■		
Improve dam operations		■		■	
Build pipeline			■	■	■
Behaviour					
Voluntary targets		■			■
Demand pricing	■	■	■		
Water restrictions or pressure reduction	■	■	■		
Supply/demand					
Flow regulators	■	■			■
Building codes		■			■
Desalination					
Communication					
Real-time dashboard	■			■	■
Outreach campaign	■	■	■		

Source: Sara Hughes, et al., 2025, RAND Center for Climate and Energy Futures

As global temperatures rise, cities throughout the US are more likely to experience hotter, drier conditions like this. Those conditions increase evaporation, drying out vegetation and soil and lowering groundwater tables.

The Northeast drought was easing in much of the region in early 2025, but communities across the US should take note of what happened. They can learn from the experiences of cities that have had to confront major water supply crises – such as Cape Town, South Africa; São Paulo, Brazil; Melbourne, Australia; Las Vegas; and New Orleans – and start planning now to avoid the worst impacts of future droughts.

Lessons from cities that have seen the worst

Our new analysis of these five cities' experiences provides lessons on how to avoid a water supply crisis or minimise the effects through proactive policies and planning.

Many cities have had to confront major water supply crises in recent years. Perhaps the most well-known example is Cape Town's 'Day Zero'.

After three years of persistent drought in the region, Cape Town officials in fall 2017 began a countdown to Day Zero – the point at which water supplies would likely run so low that water would be turned off in neighbourhoods and residents would need to fetch a daily allocation of water at public distribution points. Initially it was forecast to occur in April 2018.

Water rates were raised, and some households installed flow restrictors, which would automatically limit the amount of water that could be used. Public awareness and conservation efforts cut water consumption in half, allowing the city to push back its estimate for when Day Zero would arrive. And when the rains finally came in summer 2018, Day Zero was cancelled.

A second example is São Paulo, which similarly experienced a severe drought between 2013 and 2015. The city's reservoirs were reduced to just 5% of their capacity, and the water utility reduced the pressure in the water system to limit water use by residents.

Water pricing adjustments were used to penalise high water users and reward water conservation, and a citywide campaign sought to increase awareness and encourage conservation. As in Cape Town, the crisis ended with heavy rains in 2016. Significant investments have since been made in upgrading the city's water distribution infrastructure, preventing leaks and bringing water to the city from other river basins.

Planning ahead can reduce the harm

The experiences of Cape Town and São Paulo – and the other cities in our study – show how water supply crises can affect communities.

When major changes are made to reduce water consumption, they can affect people's daily lives and pocketbooks. Rapidly designed conservation efforts can have harmful effects on poor and vulnerable communities that may have fewer alternatives in the event of restrictions or shutoffs or lack the ability to pay higher prices for water, forcing tough choices for households between water and other necessities.

Planning ahead allows for more thoughtful policy design.

For example, Las Vegas has been grappling with drought conditions for the past two decades. During that time, the region implemented water-conservation policies that focus on incentivizing and even requiring reduced water consumption.

Since 2023, the Las Vegas Valley Water District has implemented water rates that encourage conservation and can vary with the availability of water supplies during droughts. In its first year alone, the policy saved 3 billion gallons of water and generated US$31 million in fees that can be used by programs to detect and repair leaks, among other conservation efforts. A state law now requires businesses and homeowner associations in the Las Vegas Valley to remove their decorative grass by the end of 2026.

Since 2002, per capita water use in Las Vegas has dropped by an impressive 58%.

Solutions and strategies for the future

Most of the cities we studied incorporated a variety of approaches to building water security and drought-proofing their community – from publishing real-time dashboards showing water use and availability in Cape Town to investing in desalination in Melbourne.

But we found the most important changes came from community members committing to and supporting efforts to conserve water and invest in water security, such as reducing lawn watering.

There are also longer-term actions that can help drought-proof a community, such as fixing or replacing water- and energy-intensive fixtures and structures. This includes upgrading home appliances, such as showers, dishwashers and toilets, to be more water efficient and investing in native and drought-tolerant landscaping.

Prioritising green infrastructure, such as retention ponds and bioswales, that help absorb rain when it does fall and investing in water recycling can also diversify water supplies.

Taking these steps now, ahead of the next drought, can prepare cities and lessen the pain.

3 February 2025

Is the world running out of freshwater?

As the climate crisis intensifies, it shrinks and contaminates the water we rely on. We need to act now to reduce the existing and emerging threat that this poses to our health.

By Simge Eva Dogan

How climate change drives Earth's freshwater decline and threatens health

- With 2°C warming, up to three billion people are projected to experience chronic water scarcity.
- Water scarcity impacts health in many ways, including human displacement, poverty, disease and food insecurity.
- We need immense political action, greater financial investment and a whole-of-society approach to fight climate change, and protect Earth's freshwater and people's health.

More than half of the world's population faces water scarcity for at least one month a year.

Safe water is a basic human right and essential for our health, whether we use it for drinking, food production or hygiene. But it's also a finite resource.

Only around 3% of Earth's water is freshwater, most of which is in frozen glaciers, ice caps or deep underground.

As climate change accelerates the frequency of extreme weather events and higher temperatures, our freshwater supply grows smaller and the risks to people's health greater.

It's one of many reasons why the world is running out of water. From an increasing population that demands more food and energy, to extensive use in agriculture and industry – stress on water is rising.

Water scarcity can cause human displacement and poverty. Contaminated water can transmit diseases like diarrhoea, cholera and polio. And for millions of women and children, the physical strain of walking up to 12 kilometres a day to collect water endangers their health.

How is the global water shortage affecting people's health?

In California, devastating droughts and surging temperatures have affected the water cycle. These extreme weather events have led to greater evaporation and changing rainfall patterns. The US state experienced its driest three years on record between 2019 and 2022.

Climate change is making the walk to water longer and hotter, damaging the health of communities around the world.

of the hike up Mount Kilimanjaro, Tanzania

the walk around Wembley Stadium, UK

of the trek to Machu Picchu, Peru

Sources: WaterAid https://www.wateraid.org/uk/get-involved/events/walk-for-water
Climbing Kilimanjaro https://www.climbing-kilimanjaro.com/how-long-it-takes-climb-mount-kilimanjaro/
Inca Trail Peru https://www.incatrail-peru.com/how-to-get-there-machu-picchu
Wembley Stadium https://www.wembleystadium.com/about/stadium-facts-and-features

While severe storms in 2023 filled previously depleted reservoirs, groundwater reserves remain critically low. Today, nearly one million Californians are affected by failing water systems.

Most cities have made investments to diversify their water supply and expand conservation efforts, but smaller rural communities that rely on wells are increasingly vulnerable. Many Californians rely on drinking water that may contain chemical contaminants. Drought can further increase these contaminants, while fire damage to the well equipment can add toxic chemicals to the water.

The changing water is already harming people's health. For example, following the 2018 Camp Fire in Butte County – California's most destructive wildfire in history – chemical contamination of the drinking water system resulted in health risks and restrictions on water use. In a survey of more than 200 households following the fire, 54% self-reported that at least one member in their household experienced anxiety, stress or depression in connection with securing safe water and contamination issues.

Nine thousand miles from the Californian crisis, Kenyans are facing even greater climate-related water shortages and the severe health risks that come with it.

Kenya contributes less than 0.1% of global greenhouse gas emissions annually but like many low-emitting countries, it is disproportionately affected from the effects of the climate crisis.

Extensive drought has led to extreme food insecurity. Ngawosa Eregai, a Community Health Worker in Turkana County, says: 'The biggest challenge we have is lack of water.'

Four or five people have to use their hands to dig deep holes in the ground where the rivers used to be to get to water. 'We drink this water because we do not have a choice,' says Ngawasa. 'We trust that it will not be harmful.'

For Esther Elaar, who fetches and carries water for her family to use every day, the four-hour walk weighs on her physical health. 'My whole body feels sore,' she says. 'I usually carry 20 litres of water at a time because the water point is far off.'

The long and hot walk to water impacts pregnant women like Esther. Their babies are dying before they're born.

'While carrying water, I feel the baby in my womb moving,' Esther says. 'A lot of women have miscarried in this area while going to look for water.'

What can be done to prevent freshwater decline?

Esther's story is just one of millions.

The groundwater in Germany which supplies almost 70% of the country's drinking water is declining. Iran's retreating Lake Urmia grows saltier as it shrinks and affects the water, soil and clean air available to local villagers. And prolonged droughts in Peru damage agriculture and access to clean water in Indigenous communities.

The climate crisis is transforming water on Earth and the health impacts reach far and wide.

With just 2°C warming, up to three billion people are projected to experience chronic water scarcity. It's projected that current policies around the world will result in around 2.7°C warming above pre-industrial levels. But we have the knowledge and tools to prevent that from happening.

It involves implementing strategies such as conserving water and restoring freshwater ecosystems like wetlands. To achieve that, we need immense political action, greater financial investment and a whole-of-society approach.

At Wellcome, we're working with the people most affected by climate change to support critical research, tools and resources. If we act collectively now against climate change, we can protect our water and our health.

16 November 2023

Why is the UK facing water shortages despite record rainfall?

Climate breakdown, an increasing population and a lack of new infrastructure are factors contributing to water scarcity.

By Helena Horton, Environment reporter

After months of deluge it seems incomprehensible that the UK could face water shortages in the event of a hot and dry spring and summer.

With an increasing population, a trend of 'all or nothing' rain caused by climate breakdown and a lack of investment in infrastructure, we do not keep the water that falls on our island, but let it wash out to sea.

Drier countries must look on in disbelief as this rainy isle lets the ever more precious resource flow down rivers and out of the country, leading to hosepipe bans and potentially more dire consequences in the summer. Here are some major reasons:

Climate

Before the impacts of climate breakdown became painfully apparent, the UK was a country that could rely on rainfall year-round. Why bother investing in water storage in a perpetually soggy country? In recent decades, this excuse has held less water as the trends have become clearer.

Jamie Hannaford, a hydrologist at the UK Centre for Ecology & Hydrology, says water shortages are frequently coming after floods: 'Summer 2022 was an exceptional drought event, which closely followed another major drought in summer 2018. It is worth noting that the intervening period saw exceptional flooding in 2019–2020 and further flooding in early 2021.

'Moreover, while relatively dry conditions continued in some areas into early 2023 following the 2022 drought, the period since mid-2023 to present has been very wet across most areas of the UK and the last autumn and winter has seen exceptional flooding. Recent years have therefore seen a pattern of hydrological "volatility".

'In fact, much of the post-2000 period has been marked by such volatility, with episodes of sustained flooding and droughts, some of which have extended the range of hydrological variability we see in the UK.'

Rivers

Before the ascent of the railway, the UK used canals for goods and transport. The golden age of the canal was between the 1770s and the 1830s. This meant that the naturally wiggly rivers, through which water flowed slowly, were straightened

and deepened for ease of transport. Meandering streams were turned into major channels, and water flows more quickly through a deep, straight river than a wiggly one spreading out over the landscape. This has implications for flooding, as water rushes from the uplands quickly to towns and villages below, and also for water capacity, as rather than being retained in rivers and wetlands it goes quickly out to sea.

Soil

We have drained our wetlands, such as the Somerset Levels, for housing and agriculture, and these used to hold a large amount of water. Over the centuries, the land has been parched to make it easier to farm and build. Agricultural land is mainly designed so any water that falls runs off into the surrounding aforementioned straightened rivers, where it is removed from the land and sluiced out to sea. The way we farm is preventing soil from holding water, too, as intensive farming of the soil strips it of its organisms and nutrients. This means instead of acting like a sponge, it is more sandy, so water falls through it and sluices off the drained land into the rivers. It also means that farmers need to abstract more water when it is dry, as they have not held it on their farms, whether that is in streams, wetlands or in the soil.

Reservoirs

Why not dig some massive reservoirs to catch all of the water that falls in autumn and winter? Good question. It has been 30 years since the last major reservoir was built in the UK – which means there has not been one since privatisation.

Some argue that there is little incentive to invest in infrastructure such as reservoirs when a private company is trying to give as much money to shareholders as possible. England is one of the few countries in the world where water is fully owned by private companies.

Water companies point to the Kafkaesque planning system and MPs who campaign ferociously about any planned infrastructure in their constituencies. Reservoir plans have been delayed or scrapped due to the political difficulties of building infrastructure in this country.

In the Environment Agency's plans, the companies commit to build seven major reservoirs between now and 2050 but many think that is not adequate.

Leaks

It is common during the baking summer months to see a geyser spouting from the road, sending gallons of water gushing down the drain due to a burst pipe. It sticks in the craw when one has been told to cut down on water use or banned from watering the garden.

Pipes leak about a trillion litres of water a year; once again this problem is the result of a lack of investment. Many pipes are Victorian and the rate of replacing them is so slow it would take 2,000 years to get them up to scratch at the current rate of repairs.

Usage

While it is imperative the government and water companies secure water supply for the future by tackling the above problems, the public has a part to play, too. The average person in the UK uses 150 litres a day. This is more than many European countries, including France, where the average is 128, Spain (130), and Germany (122). Using water butts and having toilet systems that do not use fresh tap water are ways that homes could use water more sustainably. Water UK, the body representing the water industry, says most British people shower for seven or eight minutes a day, and has called for people to shorten their showers by a couple of minutes.

1 April 2024

Questions

1. Why is the UK experiencing water shortages despite having periods of heavy rainfall?
2. How does climate change contribute to the UK's water scarcity problems?
3. What role does infrastructure play in the UK's water management, and how has this affected water availability?
4. In what ways does population growth impact water resources in the UK?
5. What are some consequences of water shortages for people and the environment?

Design a water conservation plan

Design a water conservation plan for your school, suggesting practical steps students and staff can take to reduce water usage.

Campaign

Develop a campaign to raise awareness about water scarcity in the UK, using social media, school newsletters, or presentations to share your message.

www.theguardian.com

The UK is surprisingly short of water – but more reservoirs aren't the answer

An article from The Conversation.

By Kevin Grecksch, Departmental Lecturer and Course Director MSc in Water Science, Policy and Management, University of Oxford and Kirsty Holstead, Postdoctoral Researcher, AQUACONNECT Project, Wageningen University

Despite its rainy reputation, the UK's public water supplies are often threatened by drought and water scarcity. Shouldn't the country do a better job of capturing and using all its rain?

In 2023, the government's official National Infrastructure Commission concluded that the UK needs more reservoirs. It lamented that no new reservoir has been built for more than 30 years and demanded new water infrastructure for at least 1,300 megalitres per day, the equivalent of 5,200 Olympic-sized swimming pools.

But new reservoirs are not really the best and certainly not the only answer to water supply issues. A stronger focus on reusing water and demanding less of it in the first place provide a more suitable answer.

Reservoirs typically come in two forms: either as storage reservoirs above or below ground, or as reservoir lakes with a dam which is often used to generate hydropower. Many of Britain's reservoirs were built during Victorian times and, for example, the city of Birmingham completely relies on water delivered from the Elan Valley reservoirs in Wales some 75 miles away. The vast Loch Katrine – a lake with a dam – has been the main fresh water supply for Glasgow since 1859.

Across England's private water suppliers, there are plans for ten new reservoirs, one enlargement and multiple large pipelines to transfer water within and between different regions and water companies.

Whether these will ever be built is questionable. Building a new reservoir is not only costly and highly regulated, but local opposition is fierce. Thames Water has been trying to build a new storage reservoir south of Oxford for decades, for instance, despite objections from the local council.

Reuse and reduce

The UK should look beyond reservoirs to things like water reuse. This might include catching and filtering rainwater for instance, or cleaning water typically seen as waste. Reused water reduces the reliance on surface and groundwater supplies and could significantly address water scarcity.

The success and uptake of water reuse is evident in major regions and cities such as California, Sydney, Melbourne and Beijing. Although water reuse is developing, at the same time, the developing interest is not matched in uniform practice – for instance, according to a 2017 European Commission report, only 0.08% of urban wastewater was reused in the UK compared to 97% in Cyprus.

Reducing water demand in the first place is just as important. A report on Scotland's climate adaptation plans says citizens need to reduce their water consumption, while the UK's official climate change advisory body says more must be done in terms of reducing demand than on the supply side.

Importantly, demand management is not only for you in your home. Industry and especially agriculture have a role to play too – a large majority of the UK's water use goes to industrial processes, watering crops and drinking water for animals.

Politically unpalatable

Given Britain's per person water consumption (excluding agriculture) is among the highest in Europe, the question arises why there is not more focus on demand management.

Partly, this is because, in England at least, water companies may favour building hard infrastructure as it creates assets and thereby value for shareholders.

But it's also because it is considered politically unpalatable to ask people to save water, especially in countries such as the UK where water is perceived to be abundant. Becoming a water-sensitive nation is hard when people so often look out the window and see rain.

We know it is possible, however. Similarly rainy countries such as Denmark use less water through a combination of public campaigns, water metering, and demand reduction technologies including water saving taps, toilets and showers. The UK could also change its building regulations so that water saving devices become compulsory in new homes.

Telling customers to save more water isn't the easiest job for an industry whose reputation is at its lowest level ever, amid the sewage scandal and reports of high profits and CEO bonuses. Yet it is worth it as conversations with customers can re-establish and build trust. It can also provide customers with a better appreciation of the very resource we all need to survive.

13 January 2025

What is water security and how is it impacted by climate change?

By Declan Conway

What is water security?

Water is essential to humans' everyday existence and for a range of economic activities in rural, urban and industrial contexts. Yet access to water varies dramatically, from simply turning on a nearby tap to needing to walk, queue for and carry water, sometimes over long distances. Water can also be a threat: poor water quality can make people seriously ill, while floods are among the most damaging and deadly climate-related hazards.

The Sustainable Development Goal on water articulates the global aspiration to 'ensure availability and sustainable management of water and sanitation for all'. However, this goal is far from being achieved and 771 million people globally lack basic access to clean drinking water.

All of these aspects of water are encompassed in the concept of water security. Definitions generally include the need for a reliable supply of clean, safe water in the context of variability that is being exacerbated by climate change (e.g., changes to precipitation patterns), problems with water supply systems, and competition over access to water. Water security also means an acceptable level of protection against flooding (or water-related hazards more broadly) and consideration of water's role in supporting healthy ecosystems such as maintaining water levels for wildlife in wetlands habitats.

An essential feature of water security is the recognition of profound interconnections between the physical (hydrological) and social systems around water resources – in other words, that human activities both shape and are shaped by water systems. Addressing these physical and human dimensions together is vital to making progress on water security as safe amounts of good quality water need to be present and reliably accessible to all. Accessibility requires effective provision, allocation, management and pricing of water resources – all of which are strongly influenced by social and political values.

Maintaining and improving water security represents a set of major, cross-cutting challenges. This was recognised at the 2023 UN Water Conference: 'Adding to the pressure, more than 90% of disasters are water-related, with climate change hitting hardest through water.' However, making climate-resilient progress on water security to secure safe water for all requires sustained political and public attention, alongside more finance and approaches to water management that consider the needs of all types of users.

What does climate change mean for the water cycle and water security?

Climate change is occurring as a complex mix of slow onset trends such as warming, changes in variability or fluctuations in conditions (whereby places become on average wetter or drier), and changes in the frequency and intensity of extreme events. Many of the most significant environmental and societal effects are likely to manifest through impacts on the hydrological (water) cycle and, in turn, on water security.

Changes in the timing, amount, intensity and state of precipitation (as snowfall or rainfall), and in atmospheric evaporative demand (the potential loss of water from the Earth's surface, driven by atmospheric factors), will alter the availability of surface and groundwater sources. The characteristics of the land's surface, including land use and land cover – which may change in response to higher temperatures and human activity – will also affect changes to the water cycle. The effects of glacier and snow cover shrinkage in mountain environments reflect complex interactions between water resource systems and glacial meltwater. Demand for water and future water security will be determined by a range of influences, both related to climate change (e.g., for irrigation) and not (e.g., population growth and urbanisation).

While there is very high confidence that climate change is ongoing and set to continue, there is less certainty around how exactly key elements of the water cycle will evolve. Uncertainty is particularly high in relation to future precipitation patterns, due largely to differences between climate model results and exacerbated by the fact that precipitation may fluctuate over the course of decades, inducing wet and dry periods in the same location over time. Overall, however, there will be a higher frequency and intensity of floods. It is also expected that precipitation globally will increase because warmer temperatures increase evaporation and allow the atmosphere to hold greater volumes of water vapour.

Recent events are highlighting the vulnerability of water supply systems to prolonged drying, for example the 'Millennium drought' in Australia, and drought conditions in California. Cape Town's high profile 'Day Zero' brought the city close to a major water supply crisis as reservoirs were almost depleted. While drought was not the only cause, the Cape Town case exemplifies the kind of threat that climate change represents for water security.

The implications of climate change for water security range from large-scale shifts in the flows of transboundary river systems (that cross multiple territories) to disruptions to local unimproved water sources (which include unprotected wells, springs and surface water). Changing patterns of extremes will also increase flood risk with consequences for individual and community wellbeing through physical impacts and the psychological effects of risk exposure and trauma recovery.

Wide-ranging secondary effects of climate change on food and energy production and land use are likely to cause even further disruption to water management systems.

How are water resources managed?

Management of water resources underpins urban water supply, provision of agricultural irrigation (which is globally by far the largest consumer of water) and industrial processes such as the cooling of power plants. Traditionally, management has relied on the assumption that our climate and water sources are stable (or 'stationary') over the long term, but climate change undermines this assumption.

While water management systems already have strategies to deal with a certain level of variability and uncertainty (e.g., water storage, water demand management and early warning systems), it is questionable whether these will be sufficient to cope with the projected rate and magnitude of change to water resources under climate change.

Investments in large-scale water infrastructure are still often made under the assumption that rainfall or streamflow patterns will resemble historical patterns (i.e., they are stationary over the long term). But with their long operational lifetimes they will be exposed to a much wider range of variability and extremes than currently experienced. This poses severe risks to performance and while guidelines for incorporating climate resilience into large-scale water projects are emerging, these are optional and rarely adopted.

What adaptation is needed?

Uncertainty about how the climate will evolve continues to have an important bearing on practical approaches to adaptation in the water sector. Currently, climate models lack the spatial and temporal detail often required for adaptation planning. This means that adaptation should factor greater levels of uncertainty into water resource planning and consider how water systems can cope with a wide range of future climate conditions, alongside other factors such as increasing urbanisation, development and population growth.

Water resource planners and managers are learning to address these new challenges while practitioners are grappling with changing risk profiles and their operational implications. Adaptation requires technical and engineering strategies (e.g., water storage, water saving devices, flood protection measures) to secure safe water supply as well as socioeconomic responses to maintain a sustainable level of demand and equitable accessibility. Early warning systems for flood or drought events may integrate physical (improved forecasts) and social elements (improved risk communication). Involving communities and stakeholders at all points in these processes can guide effective and robust adaptation pathways that are tailored to local contexts.

Factoring climate change into planning is already becoming a requirement in some jurisdictions. For example, the UK's Climate Change Act requires Water Resource Management Plans which are to be published every five years covering a minimum period of 25 years. Adaptation options proposed for the UK include achieving formal international standards of resilience, strengthening building regulations, water metering and enhanced drought management.

While there are many adaptation strategies to strengthen resilience and water security, challenges include unequal access to certain options and a lack of detailed evidence on their effectiveness, particularly at higher levels of global warming.

7 August 2023

Being water wise

Water is essential for life. We drink it, use it for cooking, cleaning, and even for creating the food we eat. Yet, despite its importance, we often take water for granted. In many parts of the world, water scarcity is a daily reality, and even in places where water is plentiful, wastefulness can still have negative impacts. As teenagers, becoming more aware of water conservation can make a big difference – both for our environment and our future.

Why should we be water wise?

Water scarcity is a growing problem around the world. Climate change, pollution, and population growth are all putting increased pressure on our water supplies. Even in the UK, where rain is common, there can be shortages during dry spells. Being water wise means recognising the value of water and making conscious choices to use it responsibly.

Wasting water not only affects the environment but also has a social and economic impact. As water becomes scarcer, the cost of providing clean, safe water increases. This can affect families and communities, especially those already struggling financially. On a global scale, conserving water helps reduce the energy needed to pump, treat, and distribute it – lowering our carbon footprint.

What can we do at home to save water?

Small changes in our everyday routines can make a big difference. Here are some practical and easy-to-follow tips for conserving water at home:

- Turn off the tap: Whether you're brushing your teeth or washing your hands, leaving the tap running can waste up to six litres of water per minute. Turn it off while scrubbing or brushing and only use it when you need to rinse.
- Take shorter showers: A long, hot shower can feel great, but reducing your shower time by just a few minutes can save significant amounts of water. Try setting a timer to keep your showers under five minutes.
- Fix leaking taps: Even a slow-dripping tap can waste thousands of litres of water over a year. If you notice a leak, report it to an adult or help fix it yourself.
- Use a bowl when washing up: Instead of leaving the tap running when washing dishes, use a bowl of soapy water to clean items and then rinse them quickly. This method can halve the amount of water used.
- Fill the washing machine: Only run the washing machine or dishwasher when you have a full load. This maximises water efficiency and saves energy too.
- Water plants wisely: If you have a garden, water plants early in the morning or late in the evening to reduce evaporation. Collecting rainwater in a water butt can also be a sustainable way to keep your plants hydrated.
- Re-use water where possible: Water used for rinsing vegetables can be saved to water houseplants. Similarly, if you have a pet, leftover water from their bowl can be used to water outdoor plants rather than throwing it away.

Being water wise for a sustainable future

Taking simple steps to conserve water at home not only benefits the environment but also helps reduce costs for households. It might seem like small changes don't make a difference, but when many people take action, the impact is huge. As teenagers, we have the power to form habits that will benefit us and our planet in the long run. Let's be water wise – because every drop counts.

13 best ways to save water

97.5% of the world's water is locked in seas and oceans, too salty for human use. And most of the remaining 2.5% is in the ice caps.

So we humans depend on the tiny bit available as fresh water – an essential natural resource for life. That's why we need to ensure water is kept free from pollutants and save water where we can.

But we don't just use water for drinking. We wash in it, clean with it, and use it to produce everything from clothing to food. Crop production – including feed for livestock and biofuels – is putting a great strain on fresh water supplies, and sewage and agricultural run-off are plaguing our rivers and lakes.

Save water, save our planet

Our water comes from rivers, lakes and reservoirs.

Climate breakdown is one of the biggest drivers of water shortages and is expected to heap even more pressure on our depleting supplies. People in England are facing shortages by 2050 unless we save water fast – according to the UK Environment Agency.

What's more, pollutants like sewage and farm run-off are causing havoc in rivers and lakes, like Lake Windermere in Cumbria. In 2023 alone, 440,000 hours of sewage was released along England's coastline.

Our top tips on how to save water

1. Turn off the taps

Don't let your water consumption run out of control. Save 6 litres of water a minute by turning off your tap while you brush your teeth. Fix leaky taps too – and stop what could be 60 litres of water going straight down the drain every week.

2. Shower with less

Every minute you spend in a power shower uses up to 17 litres of water. Set a timer on your phone to keep your showers short, sweet and water-saving.

Switching to an efficient shower head will allow you to lather up in less water, which means you'll save water and cut your bills.

3. Save up your dirty clothes

Washing a full machine load of clothes uses less water and energy than two half-loads . This means lower bills as well as saving water.

4. Call for a new law

We can each do our small bit to save water at home, but we need to be able to hold polluters accountable for damaging our rivers, lakes and seas. In 2023, sewage was poured into the UK's waterways 1,000 times a day.

Friends of the Earth is calling for a new law that would force polluters to clean up their act: the Right to a Healthy Environment.

5. Time your gardening

Water outdoor plants in the early morning or at the end of the day to stop water immediately evaporating in sunlight and heat. Water the soil so that the liquid goes straight to the roots, where it's needed.

In a heatwave, animals need water too. Instead of watering your lawn, leave out a water-filled container, like a casserole dish, for birds to drink from and wash. Thirsty bees and other insects will need a saucer or bowl with water and stones in it.

6. Catch rainwater

Installing water butts saves up to 5,000 litres of water a year . And your plants will thank you for rainwater rather than treated tap water.

You can also cut water use by 33% by watering plants manually instead of using automatic sprinklers.

7. Get a low-flush toilet

The average UK household uses about 30% of all water used to flush the toilet . Modern dual-flush systems save huge amounts of water. They use just 6 litres – or 4 with a reduced flush – much less than the 13 litres for each old-style single flush.

If you can't invest in a new loo, get a water-saving bag for your old-style toilet .

8. No more washing up

Got a dishwasher? Fill it up completely each time you run it and you'll use less water than you would doing the dishes by hand. Yes, even if you're using a washing-up bowl. What better excuse to go and have a nice sit down?

9. Reduce food waste

It takes a lot of water to produce our cereal, fruit and other food.

More than half of the 7 million tonnes of food and drink UK households bin every year could be eaten. Wasting less food could not only save water, but also save you £540 a year.

10. Boil what you need

Save water, money and energy by only boiling as many cups of water as you need.

11. Steam your veggies

Steam your food to cut water usage and retain more of the natural nutrients.

If you do boil, try using the leftover water as a tasty stock for soups. Or let it cool and use it to water plants.

12. Be plumbing prepared

Regularly check your kitchen pipes and the dishwasher hose for slow leaks.

Find out where your household stop valve is and make sure that you can turn it on and off. You'll thank us if you ever have a burst pipe, because you'll be able to cut off the flow before it floods the house.

Find a contact number for your water supplier , so that if you find a leak you can let them know and stop water being wasted.

13. Quality and seasonal eating

Rearing animals for meat and dairy and harvesting crops like avocado at a large and unsustainable scale is incredibly water-intensive. By cutting down on meat and dairy and eating seasonal vegetables you'll be helping to save water.

You'll also be helping to protect the climate. The meat and dairy industries are big contributors to global warming

Design

Design a poster with tips on saving water. You can use the tips in this article, or come up with your own.

Family challenge: the one-week water-saving challenge

Develop a one-week plan for your household to reduce water usage. Your plan should include:

- A list of daily tasks or changes to routine (e.g., using a shower timer, collecting rainwater for plants).
- A tracking sheet to record daily water usage before and after implementing the changes.
- A reflection journal entry at the end of the week, summarising what worked well, what was challenging, and how much water was saved.

Share your findings with the class, including any changes your family plans to keep.

A timer can shorten your shower even when you have no incentive to save water – new study

An article from The Conversation.

By Pablo Pereira-Doel, Water Literacy and Sustainable Water Behaviour area lead at the Institute for Sustainability, University of Surrey and Xavier Font, Professor of Sustainability Marketing, University of Surrey

From Barcelona to Mexico City, popular tourist destinations are facing droughts and running out of water. Yet, a huge 40% of household water is used for showers and baths, with products like shower gel actively encouraging people to spend longer washing, so finding simple and scalable ways to cut water consumption is crucial.

Our team of social scientists installed smart sensors in shower cubicles to test how best to encourage people to save water while showering. This timer device detects when a shower is taking place and updates the person about how long they've been showering.

Our new study of more than 17,500 showers from six hotels and youth hostels in Denmark, Spain and the UK compared ways to convince holidaymakers, who had zero incentive to shorten their showers, to do just that.

We found that regular showers (monitored with hidden timers) lasted an average of 297 seconds, or nearly five minutes, and used 38 litres of water. But smart showers that displayed the shower duration lasted 220 seconds. That's 25.79% shorter, saving the equivalent of about 10 litres per shower.

Installing smart technology that provides this continuous, real-time feedback during showers reduces showering time. These timers caught and maintained people's attention, made them more aware of their water usage and motivated them to take shorter showers.

If people with no incentive can shorten their showers simply by installing a timer that reminds them how long they are spending there, then we expect that the savings in households could be even greater.

In another study, co-authored with sustainable water use expert James Daly and environmental psychologist Ian Walker, we used the same methods to measure more than 86,000 showers for 39 weeks in UK university student halls of residence.

This research found that with middling water pressures our smart device reduced water consumption by up to 43%, saving 25 litres of water per shower. Surprisingly, reducing water pressure doesn't help save water. Higher water pressure leads to shorter showers and less water consumption overall, with and without the sensor. That's because pressure interacts with the flow, so the best combination to reduce water consumption is high pressure and low-flow showers.

Based on our studies of tourists and university students, we found that this smart technology was effective in different contexts and over a long time with the same users.

The tourist water conundrum

Tourists are not financially motivated to save resources because they don't have to pay separate water bills when staying at a hotel. They don't experience peer pressure because showering is a private behaviour, and people tend not to be in a hurry while on holiday so a long shower may be more of a treat.

But Tenerife has just declared a drought emergency. Parts of Spain, southern France, most of Italy, Malta, most of Greece and other tourist destinations are also suffering from severe droughts.

All of these places welcome millions of tourists every year, usually in summertime when water availability is at its lowest. In the busiest parts of the Balearic Islands, tourism is responsible for 58% of the total water consumption. The stark reality is that residents of Tenerife consume 150 litres of water per person per day, while the 6.5 million tourists who visit every year consume between 300 and 400 litres per person per day.

From hotels to halls

The effectiveness of the smart sensor when used in hotels could potentially have been due to its novelty, so we wanted to research an alternative scenario and investigate how best to shorten showers over a long period of time. University students in UK campuses reportedly have showers of 10–12 minutes, significantly longer than the national average, and they stay on campus for several months.

After using the smart sensor for their first shower, people tend to shower for a similar amount of time for subsequent showers. This suggests people may 'anchor' their shower duration to their initial experience.

Adding persuasive messages has some benefit too, but not as much as the timer. The type of sustainability message can increase or decrease people's engagement so wording needs to be carefully designed based on the science. The most effective message in shortening showers appealed to people's personal values, and set an ambitious goal with the words: 'Water is essential for life! A typical shower here has water running for three minutes 30 seconds. Will you beat the clock? Water conservation starts with you. Make a difference.'

The shower sensors we used aren't cheap (they cost around £90 each), so why not safely prop up your phone somewhere in the bathroom and set a countdown, ending with an alarm sound, to see whether you too can beat the clock.

16 April 2024

Persuasive writing

Write an open letter aimed at tourists visiting areas with water scarcity. Your letter should:

- Explain why reducing shower time is important, particularly in areas facing drought.
- Highlight how tourists can have a significant impact on local water supplies.
- Suggest practical ways they can reduce their water usage while on holiday.
- Use persuasive language to encourage responsible behaviour.

Design a water-wise travel guide

Create a mini travel guide aimed at promoting water conservation for holidaymakers. Your guide should include:

- A section on why reducing shower time matters, particularly in regions with limited water supplies.
- Tips and tricks for using less water in hotel rooms and holiday accommodations.
- A 'Did You Know?' section with facts about water usage by tourists.
- A quiz or challenge to encourage mindful water use.

Useful Websites / Further Reading

Useful websites

www.bda.uk.com

www.friendsoftheearth.uk

www.independent.co.uk

www.lse.ac.uk

www.perfectbalanceclinic.com

www.theconversation.com

www.theguardian.com

www.waterforgood.org

www.wave-utilities.co.uk

www.wellcome.org

www.who.int

Agriculture

The practice of cultivating soil, growing crops, and raising animals for food, fibre, and other products used to sustain and enhance human life. Agriculture can include both large-scale farming and smaller, subsistence-based practices.

Chlorine

A chemical element commonly used to disinfect water, making it safe to drink by killing bacteria and viruses. Chlorine is often added to public water supplies and swimming pools to maintain water quality and prevent the spread of waterborne diseases.

Cholera

An infectious disease affecting the small intestine, usually contracted from contaminated water. It causes severe diarrhoea and vomiting, leading to dehydration. Cholera outbreaks often occur in areas with poor sanitation and limited access to clean water.

Climate

The long-term pattern of weather conditions in a particular area, including temperature, precipitation, and seasonal variations. Climate differs from weather, which refers to short-term atmospheric conditions.

Climate change

A shift in global or regional climate patterns, often linked to increased levels of atmospheric carbon dioxide. It results from natural processes or human activities, causing altered weather patterns and more frequent extreme climatic events.

Contamination

The presence of harmful or unwanted substances in the environment, making water, air, or food unsafe to use. Contamination can result from pollution, chemical spills, or improper waste disposal.

Deforestation

The large-scale removal of forests for timber, agriculture, or livestock grazing. Deforestation reduces biodiversity, contributes to climate change, and disrupts local ecosystems.

Drought

An extended period of unusually low rainfall, resulting in a shortage of water. Droughts can impact agriculture, water supplies, and ecosystems, leading to food scarcity and habitat loss.

Ecosystem

A community of living organisms interacting with each other and their physical environment. Ecosystems can range from forests and wetlands to urban areas, each with complex networks of plants, animals, and microorganisms.

Effluent

Wastewater or sewage discharged into a natural body of water from industrial processes or sewage treatment plants. Effluent can pollute water sources, impacting aquatic life and human health.

Glossary

Environment
The surroundings or conditions in which a person, animal, or plant lives. This includes physical, social, and cultural factors that influence growth, health, and wellbeing.

Estuary
A coastal area where freshwater from rivers meets and mixes with saltwater from the sea. Estuaries support diverse ecosystems and are crucial breeding grounds for many marine species.

Filtration
A process that removes impurities from liquids or gases by passing them through a filter. Filtration is commonly used to purify drinking water, removing contaminants like dirt, bacteria, and chemicals.

Flooding
The overflow of water onto normally dry land, often caused by heavy rainfall, river overflow, or storm surges. Flooding can damage property, disrupt communities, and affect ecosystems. Flood-prone areas often have flood management systems to reduce risks.

Floodplain
A flat area next to a river or stream that regularly floods. Floodplains are vital for absorbing floodwaters and maintaining natural habitats, but they can pose risks to nearby communities.

Groundwater
Water located beneath the Earth's surface in soil pore spaces and aquifers. It supplies wells and springs and is a vital resource for drinking water and irrigation.

Groundwater extraction
The process of withdrawing groundwater from aquifers faster than it can naturally replenish, leading to depletion and environmental issues like subsidence and reduced water quality.

Hosepipe ban
A temporary restriction on using hosepipes, often imposed during droughts to conserve water. Violating a hosepipe ban can result in fines and is aimed at reducing unnecessary water usage.

Hydropower (or water power)
Renewable energy generated from the movement of water, such as flowing rivers or tidal movements. Hydropower plants convert kinetic energy from water into electricity.

Irrigation
The artificial application of water to crops to aid in growth. Irrigation systems help maintain agriculture in dry regions and include methods like sprinklers, drip systems, and canals.

Ocean
A vast body of salt water that covers a large portion of the Earth's surface. Oceans regulate global climate, support marine life, and are vital for transportation and food sources.

Potable water
Water that is safe for human consumption. It has been treated or filtered to remove harmful contaminants and is free from bacteria, viruses, and pollutants.

River
A natural watercourse flowing towards an ocean, sea, lake, or another river. Rivers are vital ecosystems and are used for drinking water, irrigation, and transportation.

Runoff
Water that flows over land, often after rainfall, moving towards rivers, lakes, or oceans. Runoff can carry pollutants from urban or agricultural areas, impacting water quality.

Sea
A large body of salt water smaller than an ocean, often partly enclosed by land. Seas are vital for marine biodiversity and human activities like fishing and tourism.

Stream
A small, flowing body of water, often a tributary to a river. Streams support local ecosystems and are crucial for maintaining the hydrological cycle.

Surface water
Water found on the surface of the Earth, such as in rivers, lakes, and reservoirs. Surface water is crucial for ecosystems and human use but is vulnerable to contamination.

Sustainability
Living in a way that meets current needs without compromising future generations' ability to meet theirs. It involves managing resources responsibly and protecting the environment.

Tidal power
A form of renewable energy derived from the movement of ocean tides. Tidal energy can be harnessed using turbines to generate electricity.

Water quality
A measure of the condition of water, based on chemical, physical, and biological factors. Good water quality is essential for drinking, agriculture, and ecosystem health.

Water scarcity
A lack of sufficient water resources to meet the needs of a region. It can be caused by drought, overuse, pollution, or inadequate infrastructure.

Water supply
The provision of water to communities for drinking, sanitation, and industrial use. A reliable water supply is essential for public health and wellbeing.

Wastewater
Used or contaminated water from homes, industries, or agricultural processes. Wastewater treatment is essential to remove pollutants before discharge into the environment.

Index